Partially Homomorphic Encryption

Çetin Kaya Koç • Funda Özdemir
Zeynep Ödemiş Özger

Partially Homomorphic Encryption

Çetin Kaya Koç
Department of Computer Science
University of California
Santa Barbara, CA, USA

Funda Özdemir
Faculty of Engineering
and Natural Sciences
Istinye University
Istanbul, Turkey

Zeynep Ödemiş Özger
Faculty of Engineering and Architecture
Kâtip Çelebi University
İzmir, Turkey

ISBN 978-3-030-87631-9 ISBN 978-3-030-87629-6 (eBook)
https://doi.org/10.1007/978-3-030-87629-6

This Springer imprint is published by the registered company Springer Nature Switzerland AG
The registered company address is: Gewerbestrasse 11, 6330 Cham, Switzerland

Preface

The potential applications of homomorphic operations over encryption functions were recognized and appreciated almost about the same time as the first public-key cryptographic algorithm RSA was invented. The RSA algorithm is multiplicatively homomorphic. The ensuing 30 years have brought on several additively or multiplicatively homomorphic encryption functions with increasing algorithmic inventiveness and complicated mathematics. However, until 2009, it was not clear whether a fully homomorphic encryption algorithm, one that allows both additive and multiplicative homomorphisms, would exist. This was resolved by Craig Gentry, and followed up by several authors who have proposed fully homomorphic encryption algorithms and addressed issues related to their formulation, arithmetic, efficiency and security. While formidable efficiency barriers remain, we now have a variety of fully homomorphic encryption algorithms that can be applied to various private computation problems in healthcare, finance and national security.

This book started with the desire and motivation to describe and implement partially homomorphic encryption functions using a unified mathematical notation. We hope that we accomplished our projected goal, and you will find our exposition helpful to understand and implement these algorithms. If a highly efficient fully homomorphic encryption function for multiple applications were to become available, the usefulness of partially homomorphic encryption functions would be in question. But this lofty end is not yet in sight. Studying partially homomorphic encryption functions may help us to understand the difficulties ahead and perhaps to avoid blind alleys and dead ends. Moreover, partially homomorphic encryption algorithms may have certain perhaps limited applications for which significantly more efficient implementations can be obtained. If not, it is always enjoyable to learn interesting algorithms and their underlying beautiful mathematics.

The authors thank their families and dedicate this book to them.

Çetin Kaya Koç
Funda Özdemir
Zeynep Ödemiş Özger

Contents

Chapter 1
Introduction

Homomorphic encryption is the last chapter for now in the 3000-year history of cryptography and about 50 years of public-key cryptography. Performing meaningful computations with encrypted text without decrypting it seems magical at first. Imagine a deck of cards with unknown numbers on each, facing down on the table: Can you add these numbers and obtain the total value with 100% accuracy, without peeking at any of them? Homomorphic encryption can do this for you; so can a successful illusionist using some elaborate tricks, but homomorphic encryption is honest and straightforward, and involves no tricks whatsoever; it is just beautiful mathematics.

Answers to the question of how homomorphic cryptography accomplishes this feat are found in tens of doctoral theses and in hundreds of journal, conference and electronic archival articles. This book brings together several of these algorithms with their descriptions, algorithmic steps, examples, and security analyses. We are limiting our scope however to *partially homomorphic encryption functions*, which are those that allow one or a few arithmetic operations to be performed homomorphically, rather than all arithmetic operations. We also included two *somewhat homomorphic encryption functions*; however, focused on their partial homomorphisms only. The reader should note that, in this book, we use the terms homomorphic encryption and homomorphic cryptography interchangeably.

The general concept of homomorphic encryption was introduced by Rivest, Adleman and Dertouzous in [22] using the term *privacy homomorphism*, soon after the RSA algorithm was invented [21]. The multiplicative homomorphic property of the RSA algorithm was first noted in print in this paper, as $(x^e)(y^e) = (xy)^e$. Unfortunately, the RSA algorithm does not allow additive homomorphism, and therefore homomorphically evaluating a polynomial is not possible using the RSA encryption function.

The breakthrough in the search for fully homomorphic encryption functions came with the doctoral dissertation of Craig Gentry in 2009. He introduced the first construction for a fully homomorphic encryption scheme [8]. Furthermore, since 1978, we have seen very productive research on partial homomorphisms of existing or new encryption functions.

Ç. K. Koç et al., *Partially Homomorphic Encryption*,
https://doi.org/10.1007/978-3-030-87629-6_1

A taxonomy of fully versus partial homomorphic encryption is however too simplistic. Basic arithmetic operations (addition and multiplication) or Boolean gates (AND, OR, NOT) are just the building blocks for evaluating polynomials, arithmetic or Boolean circuits, or other atomic computations such as search for minimum or maximum, or compare and sort operations. A particular homomorphic property of an encryption function cannot be considered separately from its application to perform private computations. There are mathematical issues as well as efficiency considerations. Particularly, circuit depth issues are significant barriers to efficient computations.

This book specifically concentrates on partially homomorphic encryption functions introduced or analyzed since the RSA algorithm. We exclude the subjects of *somewhat homomorphic*, *leveled homomorphic* and *fully homomorphic* encryption functions. This selection leaves us a handful of mathematically interesting encryption functions. Our observation is that these algorithms are scattered in journals, conference proceedings and electronic archives, and even though they have much in common and they are designed using a few mathematical building blocks, a unified approach to their descriptions and analyses is missing; each one of them seems to be a separate mountain to climb. The need to unify the notation and start with a set of well-defined building blocks to describe and analyze the encryption functions is behind our motivation to write this book.

There is a unifying way we view and present these algorithms in this book. First of all, we wanted it to be more or less self-contained, so we included a pretty extensive mathematical background. Almost everything one needs beyond a four-year computer science or mathematics education in order to understand the encryption chapters is included in Chapter 2; however, this chapter also serves a secondary and more important purpose: It unifies the notation to describe the key generation, encryption, decryption, and homomorphic operations for each of the algorithms presented. We spent a great deal of our time to provide a unified mathematical background, since this was also the only way for us to understand these partially homomorphic encryption algorithms for the purpose of implementing them.

Chapter 2
Mathematical Background

In this chapter, we present the necessary mathematical background for the encryption algorithms in the subsequent chapters.

2.1 Number Theory

The set of integers is represented as $\mathbb{Z} = \{\ldots,-2,-1,0,1,2,\ldots\}$.

2.1.1 *Divisibility and Factorization*

Given two integers $a,b \in \mathbb{Z}$, we say that a **divides** b or b is divisible by a and write $a \mid b$ if there is an integer c such that $a \times c = b$. If a does not divide b, we write $a \nmid b$.
Division Algorithm. Let $a,b \in \mathbb{Z}$. Then there exist unique integers q and r such that

$$a = q \times b + r \text{ and } 0 \leq r < b,$$

where a is the dividend, b is the divisor, q is the quotient and r is the remainder. Notice that b divides a if and only if $r = 0$.

An integer $p \geq 2$ is said to be **prime** if its only positive divisors are the trivial ones: 1 and p. Thus if $a \mid p$ and p is a prime, then either $a = 1$ or $a = p$. An integer greater than 2 that is not prime is said to be **composite**. Note that 1 is neither prime nor composite. The following is an important divisibility property of primes.

Theorem 1 *Let $a,b \in \mathbb{Z}$ and p be a prime. If $p \mid (a \times b)$ then $p \mid a$ or $p \mid b$.*

Prime numbers are the irreducible multiplicative building blocks of the positive integers, as stated in the following theorem.

Theorem 2 ***(Fundamental Theorem of Arithmetic)*** *Let $n > 1$ be an integer. Then there exists a unique set of prime numbers $\{p_1, p_2, \ldots, p_r\}$ and positive integer ex-*

Ç. K. Koç et al., *Partially Homomorphic Encryption*,
https://doi.org/10.1007/978-3-030-87629-6_2

ponents $\{e_1, e_2, \ldots, e_r\}$ *such that*

$$n = p_1^{e_1} \times p_2^{e_2} \times \cdots \times p_r^{e_r}.$$

This unique decomposition of n into prime powers is called the **prime factorization** of n.

It is easy to compute the product of primes. However, it is very difficult to factor the resulting product of large distinct primes. The problem of determining the prime factorization of $n = p \times q$, where p and q are large distinct primes, is called the **Integer Factorization Problem** (or Factoring Problem\Factorization Problem), denoted by **Fact[*n*]**, on which the security of many public-key cryptosystems is based. Most cryptosystems use $n = p \times q$ and some of them use $n = p^2 \times q$, where p and q are chosen large primes having the same bit length.

2.1.2 Greatest Common Divisor (GCD)

Given two nonzero integers $a, b \in \mathbb{Z}$, we say that g is a common factor or common divisor of a and b if $g \mid a$ and $g \mid b$. We call g the **greatest common divisor** of a and b and write gcd$(\boldsymbol{a}, \boldsymbol{b})$ if g is nonnegative and all other common divisors of a and b divide g. Note that $\gcd(a, 0) = a$.

We say that a and b are **relatively prime** if $\gcd(a, b) = 1$, or equivalently that their only common divisors are ± 1. The following theorem is easy to observe.

Theorem 3 *If* $g = \gcd(a, b)$, *then* $\frac{a}{g}$ *and* $\frac{b}{g}$ *are relatively prime.*

Proposition 1 *If the integers a and b have the prime factorization*

$$a = p_1^{e_1} \times p_2^{e_2} \times \cdots \times p_r^{e_r}$$
$$b = p_1^{f_1} \times p_2^{f_2} \times \cdots \times p_r^{f_r},$$

where each exponent is greater than or equal to 0, then

$$\gcd(a, b) = p_1^{\min(e_1, f_1)} \times p_2^{\min(e_2, f_2)} \times \cdots \times p_r^{\min(e_r, f_r)}.$$

In the above proposition, zero exponents are used to make the set of primes $\{p_1, p_2, \ldots, p_r\}$ the same for both a and b.

Euclidean Algorithm. The most commonly used algorithm for computing the greatest common divisor of two integers is the Euclidean algorithm which solves the problem by successive division. This algorithm is based on the property that

$$\gcd(a, b) = \gcd(b, a - q \times b),$$

where q is the integer division $q = \lfloor a/b \rfloor$. Here the problem of finding the gcd of two given integers is reduced to that of finding the gcd of two smaller integers. This reduction rule is applied recursively until the algorithm obtains $\gcd(g,0) = g$, which is equal to $\gcd(a,b)$ since each iteration preserves the gcd of the previous iteration.

Algorithm 1 Euclidean algorithm for GCD

1: **function** GCD(a,b)
2:     **Input:** Nonnegative integers a and b with $a > b$.
3:     **Output:** $g = \gcd(a,b)$.
4:     **while** $b \neq 0$ **do**
5:         $q := \lfloor a/b \rfloor$
6:         $r := a - q \times b$
7:         $a := b$
8:         $b := r$
9:     **return** a

Example 1 *Let's use the Euclidean algorithm to calculate* $\gcd(72,27)$.

$$\begin{aligned}
\gcd(72,27) &= \gcd(27,72-2\times 27)\\
&= \gcd(27,18)\\
&= \gcd(18,27-1\times 18)\\
&= \gcd(18,9)\\
&= \gcd(9,18-2\times 9)\\
&= \gcd(9,0)\\
&= 9.
\end{aligned}$$

Another well-known algorithm to find the gcd of two nonnegative integers is the **binary GCD algorithm**, which is also known as Stein's algorithm [25]. This algorithm employs shift operations instead of expensive division operations, which makes it more efficient than the Euclidean GCD algorithm.

The following well-known theorem, which is known as Bézout's Lemma, shows that the greatest common divisor of two integers can be written as a linear combination of them.

Theorem 4 ***(Bézout's Lemma)*** *Given any two nonzero integers a and b, there exist integers s and t such that $s \times a + t \times b = \gcd(a,b)$. In particular, if a and b are relatively prime, then there exist integers s and t such that $s \times a + t \times b = 1$.*

The integers s and t are called Bézout coefficients for a and b and they are unique up to integer multiples of a and b.

Extended Euclidean Algorithm. The Bézout coefficients s and t above can be computed using the extended Euclidean algorithm (EEA), which applies the back-substitution method through the remainders in the Euclidean algorithm.

Algorithm 2 Extended Euclidean Algorithm (EEA)

```
1: function EEA(a,b)
2:     Input: Integers a and b not both zero.
3:     Output: (g,s,t) where g = gcd(a,b) = s × a + t × b.
4:     g0 := a, g1 := b
5:     s0 := 1, s1 := 0
6:     t0 := 0, t1 := 1
7:     while g1 ≠ 0 do
8:         q := ⌊g0/g1⌋
9:         (g0,g1) := (g1, g0 − q × g1)
10:        (s0,s1) := (s1, s0 − q × s1)
11:        (t0,t1) := (t1, t0 − q × t1)
12:    g := g0, s := s0, t := t0
13:    return (g, s, t)
```

Example 2 *Consider $a = 72$, $b = 27$. The above algorithm computes the following table*

iteration	q	g_0	g_1	s_0	s_1	t_0	t_1
0	-	72	27	1	0	0	1
1	2	27	18	0	1	1	-2
2	1	18	9	1	-1	-2	3
3	2	9	0	-1	3	3	-5

Therefore EEA returns $g = 9$, $s = -1$ and $t = 3$, with the property that

$$g = s \times a + t \times b.$$

Thus we have

$$9 = (-1) \times 72 + 3 \times 27.$$

2.1.3 Modular Arithmetic

Let $a, b, n \in \mathbb{Z}$ and $n > 0$. We say that a is **congruent** to b modulo n if and only if $n \mid (a - b)$, and write $a \equiv b \pmod{n}$. Here n is called the **modulus**. If $n \nmid (a - b)$, we write $a \not\equiv b \pmod{n}$, and say that a and b are **incongruent** modulo n.

Congruence modulo n is an equivalence relation, which means a binary relation that is reflexive, symmetric and transitive. That is, for all $a, b, c \in \mathbb{Z}$,

(i) $a \equiv a \pmod{n}$.
(ii) If $a \equiv b \pmod{n}$, then $b \equiv a \pmod{n}$.
(iii) If $a \equiv b \pmod{n}$ and $b \equiv c \pmod{n}$, then $a \equiv c \pmod{n}$.

The next theorem connects the congruence notion with the division algorithm.

Theorem 5 $a \equiv b \pmod{n}$ *if and only if a and b have the same remainder when divided by n. Moreover, both a and b are congruent to that common remainder.*

If we apply the division algorithm to a with divisor n, we get $a = q \times n + r$ for some integers q and r with $0 \leq r < n$. Then $q \times n = a - r$, which yields $n \mid (a - r)$, or equivalently $a \equiv r \pmod{n}$. So we have the following result.

Theorem 6 *For $n \geq 2$, every integer is congruent modulo n to exactly one element of the set* $\{0, 1, 2, \ldots, n-1\}$.

The sets of integers which consist of mutually congruent integers modulo n are called the **congruence classes** modulo n. Each class has no element in common with any other class, and every integer lies in a unique class. Separating the set of integers into these nonempty, mutually disjoint classes is known as partitioning the set of integers. Since each integer is congruent to its remainder upon division by n, there are exactly n congruence classes modulo n, one for each of the remainders $0, 1, 2, \ldots, n-1$. The set of integers modulo n is denoted by $\mathbb{Z}_n = \{0, 1, \ldots, n-1\}$. Here each element of $\mathbb{Z}_n$ can be thought of as an equivalence class.

For example, consider $\mathbb{Z}_5 = \{0, 1, 2, 3, 4\}$. Here, each element of the set represents an infinite set of negative and positive integers:

$$\{\ldots, -15, -10, -5, 0, 5, 10, 15, \ldots\}$$
$$\{\ldots, -14, -9, -4, 1, 6, 11, 16, \ldots\}$$
$$\{\ldots, -13, -8, -3, 2, 7, 12, 17, \ldots\}$$
$$\{\ldots, -12, -7, -2, 3, 8, 13, 18, \ldots\}$$
$$\{\ldots, -11, -6, -1, 4, 9, 14, 19, \ldots\}$$

Actually the elements of $\mathbb{Z}_n$ can be represented in two different ways: the **Least Positive** (LP) representation and the **Least Magnitude** (LM) representation:

- The Least Positive representation is $\mathbb{Z}_n = \{0, 1, 2, \ldots, n-1\}$.
- The Least Magnitude representation is

$$\mathbb{Z}_n = \{-(n-1)/2, \ldots, -2, -1, 0, 1, 2, \ldots, (n-1)/2\} \text{ if } n \text{ is odd,}$$

$$\left.\begin{array}{rl} \mathbb{Z}_n = & \{-n/2+1, \ldots, -2, -1, 0, 1, 2, \ldots, n/2\} \\ \text{or} & \\ \mathbb{Z}_n = & \{-n/2, \ldots, -2, -1, 0, 1, 2, \ldots, n/2-1\} \end{array}\right\} \text{if } n \text{ is even.}$$

In other words, a is called LM modulo n if $|a| \leq |n - a|$. For modular arithmetic operations, LP representation is often preferred.

Modular arithmetic in $\mathbb{Z}_n$, where operations are performed modulo n, has similar properties to arithmetic in $\mathbb{Z}$, such as commutativity, associativity, distributivity of addition and multiplication, and the existence of the identity elements 0 and 1 for addition and multiplication, respectively. Furthermore, every element has an additive inverse: The additive inverse of $a \in \mathbb{Z}_n$ is $n - a$ since $a + (n - a) = (n - a) + a = 0$.

However, there is a remarkable difference in the case of multiplicative inverse. As in $\mathbb{Z}$, 0 does not have a multiplicative inverse in $\mathbb{Z}_n$. In $\mathbb{Z}$, the only elements with a multiplicative inverse are 1 and -1. However, there can be other nonzero elements with multiplicative inverses in $\mathbb{Z}_n$ other than 1 and $n-1$ (where $n-1$ stands for -1 modulo n).

2.1.4 Modular Exponentiation

Given exponent e, base b, and modulus n, the computation of $a \equiv b^e \pmod{n}$ is the modular exponentiation operation. This operation forms the basis of almost all cryptographic algorithms and protocols. The **binary method**, which is also known as the square-and-multiply method, is the most commonly used one among the exponentiation algorithms, which are the factor method, power tree method, binary method, m-ary methods and sliding windows. The binary method uses the binary expansion of the exponent $e = (e_{k-1}e_{k-2}\cdots e_1e_0)$, and performs squaring and multiplication operations at each bit-scanning step. There are two versions of the binary method depending on how exponent bits are scanned: **Left-to-Right (LR)** and **Right-to-Left (RL)**.

Algorithm 3 LR binary method of exponentiation

```
1: function LRPOW(b, e, n)
2:     Input: Base b, exponent e = (e_{k-1}e_{k-2}···e_1e_0), and modulus n.
3:     Output: a ≡ b^e (mod n).
4:     if e_{k-1} = 1 then a := b
5:     else a := 1
6:     for i = k-2 downto 0 do                 ▷ Scan from left to right.
7:         a := a × a (mod n)                  ▷ Square.
8:         if e_i = 1 then a := a × b (mod n)  ▷ Multiply.
9:     return a
```

The following example shows how to compute a modular exponentiation using the LR method.

Example 3 *Let* $e = 25 = (11001)$ *and* $k = 5$. *Since* $e_{k-1} = 1$, *initialize:* $a = b$.
1st step: $a = b^2$; *since* $e_3 = 1$, *compute* $a = b^2 \times b = b^3$,
2nd step: $a = (b^3)^2 = b^6$; *since* $e_2 = 0$, *no multiplication,*
3rd step: $a = (b^6)^2 = b^{12}$; *since* $e_1 = 0$, *no multiplication,*
4th step: $a = (b^{12})^2 = b^{24}$; *since* $e_0 = 1$, *compute* $a = b^{24} \times b = b^{25}$.

The following example shows how to compute a modular exponentiation using the RL method.

Example 4 *Let* $e = 25 = (11001)$ *and* $k = 5$. *Let* $c = b$ *and initialize the result to* $a = 1$.

Algorithm 4 RL binary method of exponentiation

1: **function** RLPOW(b, e, n)
2: **Input:** Base b, exponent $e = (e_{k-1}e_{k-2}\cdots e_1e_0)$, and modulus n.
3: **Output:** $a \equiv b^e \pmod{n}$.
4: $a := 1$ and $c := b$
5: **for** $i = 0$ **to** $k-2$ **do** ▷ Scan from right to left.
6: **if** $e_i = 1$ **then** $a := a \times c \pmod{n}$ ▷ Multiply.
7: $c := c \times c \pmod{n}$ ▷ Square.
8: **if** $e_{k-1} = 1$ **then** $a := a \times c \pmod{n}$ ▷ Multiply.
9: **return** a

1st step: since $e_0 = 1$, *compute* $a = a \times c = 1 \times b = b$ *and square* $c = b^2$,
2nd step: since $e_1 = 0$, *just square* $c = (b^2)^2 = b^4$,
3rd step:since $e_2 = 0$, *just square* $c = (b^4)^2 = b^8$,
4th step: since $e_3 = 1$, *compute* $a = b \times b^8 = b^9$ *and square* $c = (b^8)^2 = b^{16}$,
Final step: since $e_4 = 1$, $a = b^9 \times b^{16} = b^{25}$.

2.1.5 Modular Inverse

Theorem 7 *An element* $a > 0$ *has a multiplicative inverse in* $\mathbb{Z}_n$ *if and only if* $\gcd(a,n) = 1$.

The Extended Euclidean Algorithm allows us to compute the multiplicative inverse of an integer a modulo n provided that $\gcd(a,n) = 1$. The EEA yields the equation $s \times a + t \times n = 1 = \gcd(a,n)$. After taking this equation modulo n, one obtains

$$\begin{aligned} s \times a + t \times 0 &\equiv 1 \pmod{n} \\ s \times a &\equiv 1 \pmod{n} \\ a^{-1} &\equiv s \pmod{n}. \end{aligned}$$

Algorithm 5 Multiplicative inverse via EEA

1: **function** INVERSE(a,n)
2: **Input:** Integer a and modulus n.
3: **Output:** $x \equiv a^{-1} \pmod{n}$ provided that $\gcd(a,n) = 1$.
4: EEA(a, n)
5: **if** $g = 1$ **then** $x := s$
6: **else** PRINT("inverse does not exist")

Example 5 $\gcd(23,25) = 1$ *guarantees the existence of the multiplicative inverse of 23 modulo 25. The EEA gives us the linear combination*

$$1 = 12 \times 23 - 11 \times 25.$$

Therefore, $23^{-1} \equiv 12 \pmod{25}$*, which can be easily verified as*

$$12 \times 23 = 276 \equiv 1 \pmod{25}.$$

The following theorem, whose statement was given by the great French mathematician Pierre de Fermat and whose first published proof was given by Leonhard Euler, is useful for computing the remainders modulo a prime p of large powers of integers.

Theorem 8 ***(Fermat's Little Theorem)*** *If p is prime, then*

$$a^{p-1} \equiv 1 \pmod{p}$$

for all $a \in \mathbb{Z}$ *with* $\gcd(a, p) = 1$.

Note that $p - 1$ is the smallest exponent satisfying this equivalence when p is prime.

Equivalently, if $a^{p-1} \not\equiv 1 \pmod{p}$ for some $a \in \mathbb{Z}$ with $\gcd(a, p) = 1$, then p is composite. However, the converse of Fermat's Little Theorem is not true: If $a^{n-1} \equiv 1 \pmod{n}$ for all $a \in \mathbb{Z}$ with $\gcd(a,n) = 1$, then n is not necessarily a prime. For example, $a^{560} \equiv 1 \pmod{561}$ for all a with $\gcd(a, 561) = 1$, but 561 is not a prime as $561 = 3 \times 11 \times 17$. Therefore, Fermat's Little Theorem can be used to prove that a positive integer n is composite, but it cannot be used to prove that n is prime. However, if the congruence $a^{p-1} \equiv 1 \pmod{p}$ holds for many values of a, then n is probably prime. Testing primality of a number using Fermat's Little Theorem is called the **Fermat primality test**.

To be able to determine whether a given number is prime without decomposing it into prime factors is possible by primality tests. Among these tests, there is a brute-force algorithm known as **trial division**: If n is not divisible by any of the primes not exceeding $\sqrt{n}$, then n is a prime number. The pseudocode below accomplishes this.

Algorithm 6 Primality test by trial division

```
1: function ISPRIME(n)
2:     Input: Integer n > 1.
3:     Output: true or false.
4:     ISPRIME := true
5:     i := 1
6:     while ISPRIME and i ≤ √n do
7:         if n = 0 (mod i) then ISPRIME := false
8:         else i := i + 1
9:     return ISPRIME
```

Fermat's Little Theorem can also be used to compute the multiplicative inverse of an element in $\mathbb{Z}_p$, which is relatively prime to p, as

$$a^{-1} \equiv a^{p-2} \pmod{p},$$

since $a^{-1} \times a \equiv a^{p-2} \times a \equiv a^{p-1} \equiv 1 \pmod{p}$.

2.1.6 Euler's Theorem

The set of positive integers that are less than n and relatively prime to n are represented as

$$\mathbb{Z}_n^* = \{a \in \mathbb{Z}_n : \gcd(a,n) = 1\}.$$

In other words, $\mathbb{Z}_n^*$ is the set of invertible elements modulo n. When $n = p$ is a prime, then $\mathbb{Z}_p^* = \{1, 2, \ldots, p-1\}$. The **Euler's phi (totient) function** is defined as the number of elements in $\mathbb{Z}_n^*$ and denoted by $\phi(n) = |\mathbb{Z}_n^*|$. Since when $n = p$ is prime, each of the numbers $1, 2, \ldots, p-1$ is relatively prime to p, one has $\phi(p) = p-1$.

The next theorem gives the main properties of Euler's phi function.

Theorem 9 *Let p be a prime and e be a positive integer. Then,*

$$\phi(p^e) = p^e - p^{e-1} = p^{e-1} \times (p-1).$$

Furthermore:

(i) Euler's phi function is multiplicative. That is, if $\{n_1, \ldots, n_r\}$ is a family of pairwise relatively prime positive integers, then

$$\phi(\prod_{i=1}^{r} n_i) = \prod_{i=1}^{r} \phi(n_i).$$

(ii) If n has the prime factorization

$$n = p_1^{e_1} \times p_2^{e_2} \times \cdots \times p_r^{e_r},$$

where the p_is are distinct prime numbers and the e_is are positive integers, then

$$\phi(n) = n \times \prod_{i=1}^{r} \left(1 - \frac{1}{p_i}\right).$$

Theorem 10 *The computation of $\phi(n)$ is polynomially equivalent to the factorization of n.*

Proof. If one knows the factorization of n, then $\phi(n)$ is calculated easily. Now let $n = p \times q$ where p and q are distinct primes, and assume that n and $\phi(n)$ are known. Since $\phi(n) = (p-1) \times (q-1) = p \times q - p - q + 1 = n - p - q + 1$, we can write $n -$

$\phi(n)+1 = p+q$. Thus we know $p\times q$ and $p+q$. Consider the following quadratic polynomial $f(x)$:

$$\begin{aligned} f(x) &= x^2-(n-\phi(n)+1)x+n \\ &= x^2-(p+q)x+p\times q \\ &= (x-p)\times(x-q), \end{aligned}$$

where the roots are p and q. Since all the coefficients of $f(x)$ are known, one can easily compute the roots, namely p and q, as:

$$\frac{1}{2}(n-\phi(n))+1\pm\sqrt{(n-\phi(n)+1)^2-4\times n}.$$

A generalization of Fermat's Little Theorem to any modulus n, which is not necessarily a prime, is Euler's Theorem.

Theorem 11 ***(Euler's Theorem)*** *Let n be a positive integer and $a\in\mathbb{Z}_n^*$. Then,*

$$a^{\phi(n)} \equiv 1 \pmod{n}$$

If p is a prime, plugging the value $\phi(p) = p-1$ into Euler's Theorem gives us the equivalence $a^{\phi(p)} = a^{p-1} \equiv 1 \pmod{p}$, which is exactly Fermat's Little Theorem. Euler's Theorem can be used to compute the multiplicative inverse for any modulus

$$a^{-1} \equiv a^{\phi(n)-1} \pmod{n}$$

for which the computation of $\phi(n)$, or equivalently the factorization of n, is required. For example, to compute $23^{-1} \bmod 25$, we need $\phi(25) = \phi(5^2) = 5^2-5^1 = 20$. Then we have $23^{-1} = 23^{20-1} = 23^{19} = 12 \pmod{25}$.

2.1.7 Carmichael's Theorem

The smallest positive divisor of Euler's phi function that satisfies the conclusion of Euler's Theorem is called **Carmichael's lambda function** and denoted by $\lambda(n)$.

Theorem 12 ***(Carmichael's Theorem)*** *Let n be a positive integer. Then $\lambda(n)$ is the smallest power satisfying*

$$a^{\lambda(n)} \equiv 1 \pmod{n}$$

for all $a\in\mathbb{Z}_n^*$.

It is trivial that $\lambda(n)\mid\phi(n)$ for all positive integers n.

Given two nonzero integers $a,b\in\mathbb{Z}$, the **least common multiple** of a and b, denoted by $\mathrm{lcm}\,(\boldsymbol{a},\boldsymbol{b})$, is the smallest positive integer which is divisible by both a and b. Since division of integers by zero is undefined, this definition is meaningful only if a and a are both different from zero. This definition can easily be extended for more than two integers.

The following theorem, which is an analogue of Theorem 9, gives the properties of the Carmichael λ function.

Theorem 13 *(i) Let p be a prime and e be a positive integer. Then*

$$\lambda(p^e) = \begin{cases} \phi(p^e) & \text{if } e \leq 2 \text{ or } p \geq 3 \\ \frac{1}{2} \times \phi(p^e) & \text{if } e \geq 3 \text{ and } p = 2. \end{cases}$$

(ii) The Carmichael λ function is not multiplicative. If $\{n_1, \ldots, n_r\}$ is a family of pairwise relatively prime positive integers, then

$$\lambda(\prod_{i=1}^{r} n_i) = \text{lcm}(\lambda(n_1), \ldots, \lambda(n_r)).$$

(iii) If n has the prime factorization

$$n = p_1^{e_1} \times p_2^{e_2} \times \cdots \times p_r^{e_r},$$

where the p_i are distinct prime numbers and the e_i are positive integers, then

$$\lambda(n) = \text{lcm}(\lambda(p_1^{e_1}), \ldots, \lambda(p_r^{e_r})).$$

Recall that there are some composite numbers n which cannot be proven to be composite by the Fermat primality test, i.e. there exist composite integers n such that $a^{n-1} \equiv 1 \pmod{n}$ for all $a \in \mathbb{Z}_n^*$. Such a composite number n is called a **Carmichael number**. It is easy to observe that n is a Carmichael number if and only if $\lambda(n) \mid (n-1)$. For example, $\lambda(561) = \text{lcm}(\lambda(3), \lambda(11), \lambda(17)) = \text{lcm}(2, 10, 16) = 80$, which divides 561-1=560. So 561 is a Carmichael number. In fact, it is the smallest Carmichael number.

2.1.8 Generators

Given an integer n and $a \in \mathbb{Z}_n^*$, the **multiplicative order** of a is the smallest exponent $e > 1$ such that

$$a^e \equiv 1 \pmod{n}.$$

The order of a is usually denoted by $\text{ord}_n(a)$, or simply $\text{ord}(a)$ if the modulus is clear from the context. It is easy to observe that for every $i \in \mathbb{Z}$, $a^i \equiv 1 \pmod{n}$ if and only if $\text{ord}(a) \mid i$. In particular, $\text{ord}(a) \mid \phi(n)$ by Euler's Theorem. An element $a \in \mathbb{Z}_n$ is called a **generator** of $\mathbb{Z}_n^*$ if $\text{ord}(a) = \phi(n)$.

Theorem 14 *If a is a generator of $\mathbb{Z}_n^*$, then $b \equiv a^i \pmod{n}$ is also a generator of $\mathbb{Z}_n^*$ if and only if $\gcd(i, \phi(n)) = 1$.*

2.1.9 Chinese Remainder Theorem

Theorem 15 ***(Chinese Remainder Theorem)*** *If $n_1, \ldots, n_k$ are pairwise relatively prime moduli, then given any integers $r_1, \ldots, r_k$, the set of linear congruences*

$$x \equiv r_i \pmod{n_i}$$

for $i = 1, \ldots, k$ has a unique solution modulo $n = n_1 \times \cdots \times n_k$, and the solution is

$$\sum_{i=1}^{k} r_i \times c_i \times m_i \pmod{n},$$

where $m_i = n/n_i$ and $c_i = m_i^{-1} \pmod{n_i}$.

Note that each modulus n_i does not need to be prime, but they need to be pairwise relatively prime. The computation of x using the linear summation formula above is called the Chinese Remainder Algorithm (CRA).

Algorithm 7 Chinese Remainder Algorithm

1: **function** CRA$(r_1, r_2, \ldots, r_k; n_1, n_2, \ldots, n_k)$
2: **Input:** Remainders $(r_1, r_2, \ldots, r_k)$ with respect to the moduli $(n_1, n_2, \ldots, n_k)$.
3: **Output:** x.
4: $n := 1$
5: **for** $i = 1$ **to** k **do**
6: $n := n \times n_i$
7: **for** $i = 1$ **to** k **do**
8: $m_i := n/n_i$
9: $c_i := m_i^{-1} \pmod{n_i}$
10: $x := 0$
11: **for** $i = 1$ **to** k **do**
12: $x := x + r_i \times c_i \times m_i$
13: $x := x \pmod{n}$
14: **return** x

Example 6 *Consider finding x modulo $n = 16170 = 15 \times 22 \times 49$ such that*

$$x \equiv 5 \pmod{15},$$
$$x \equiv 12 \pmod{22},$$
$$x \equiv 2 \pmod{49}.$$

Note that $\gcd(15,22) = 1$, $\gcd(22,49) = 1$, $\gcd(15,49) = 1$. *Here $n_1 = 15$, $n_2 = 22$, $n_3 = 49$, $r_1 = 5$, $r_2 = 12$, $r_3 = 2$. Then*

$$m_1 = n_2 \times n_3 = 1078,$$
$$m_2 = n_1 \times n_3 = 735,$$
$$m_3 = n_1 \times n_2 = 330,$$

and

$$m_1^{-1} \equiv 7 \pmod{15},$$
$$m_2^{-1} \equiv 5 \pmod{22},$$
$$m_3^{-1} \equiv 15 \pmod{49}.$$

Hence $x \equiv 5 \times 7 \times 1078 + 12 \times 5 \times 735 + 2 \times 15 \times 330 \equiv 10880 \pmod{16170}$.

2.1.10 Quadratic Residues

An integer $a \in \mathbb{Z}_n^*$ is called a **quadratic residue** modulo n if there exists an integer $x \in \mathbb{Z}_n^*$ such that

$$a \equiv x^2 \pmod{n}.$$

If there is no solution to this congruence, then a is called a **quadratic non-residue** modulo n.

Example 7 *Let* $p = 11$. *The squares of all elements in* $\mathbb{Z}_{11}^*$ *are listed in the following table.*

x	1	2	3	4	5	6	7	8	9	10
x^2	1	4	9	5	3	3	5	9	4	1

$1^2 \equiv 10^2 \equiv 1 \pmod{11}$, $2^2 \equiv 9^2 \equiv 4 \pmod{11}$, $3^2 \equiv 8^2 \equiv 9 \pmod{11}$, $4^2 \equiv 7^2 \equiv 5 \pmod{11}$, $5^2 \equiv 6^2 \equiv 3 \pmod{11}$. *Hence,* $\{1,3,4,5,9\}$ *is the set of quadratic residues and* $\{2,6,7,8,10\}$ *is the set of quadratic non-residues.*

Depending on whether the modulus n is composite or prime, we face with problems of different complexity. Let's first consider the case where the modulus is prime. The next lemma is simple and easy to verify.

Lemma 1 *For an odd prime* p *and* $a \in \mathbb{Z}_p^*$, *the congruence* $a \equiv x^2 \pmod{p}$ *has either no solution or exactly two incongruent solutions modulo* p, *namely* x *and* $-x$.

It is easy to find a nonzero quadratic residue which is $1 = 1^2$. However, it is less straightforward to find a non-residue. To decide whether a given number is a quadratic non-residue modulo p, one way is to compute all the quadratic residues modulo p and exhaustively search for a number which is not in that list. But this is not efficient. There is a useful criterion due to Euler for deciding whether an integer is a quadratic residue modulo a prime.

Theorem 16 ***(Euler's Criterion)*** *Let p be an odd prime and $a \in \mathbb{Z}_p^*$. Then a is a quadratic residue modulo p if*

$$a^{(p-1)/2} \equiv 1 \pmod{p},$$

and a is a quadratic non-residue modulo p if

$$a^{(p-1)/2} \equiv -1 \pmod{p}.$$

Corollary 1 *For an odd prime p, there are exactly $(p-1)/2$ distinct quadratic residues of p and $(p-1)/2$ distinct quadratic non-residues of p among the integers $1, 2, \ldots, p-1$.*

This corollary gives us a randomized polynomial-time algorithm for finding a quadratic non-residue: Since half of the numbers in $\mathbb{Z}_p^*$ are quadratic non-residues, a randomly chosen number from $\mathbb{Z}_p^*$ will be a non-residue with probability one half.

There is a special computational tool associated with quadratic residues, the **Legendre symbol**, denoted by $\left(\frac{a}{p}\right)$, where a is an integer and p is a prime, and defined as follows:

$$\left(\frac{a}{p}\right) = \begin{cases} 0, & \text{if } p \text{ divides } a \\ 1, & \text{if } a \text{ is a quadratic residue modulo } p \\ -1, & \text{if } a \text{ is a quadratic non-residue modulo } p. \end{cases}$$

The Legendre symbol can be seen as a function from the set of prime numbers to the set $\{0, 1, -1\}$.

Theorem 17 *Let $p \neq q$ be odd primes and a be an integer. The Legendre symbol satisfies the following properties:*

(i) $\left(\frac{a}{p}\right) \equiv a^{(p-1)/2} \pmod{p}$. *In particular,* $\left(\frac{1}{p}\right) = 1$, $\left(\frac{0}{p}\right) = 0$, *and*

$$\left(\frac{-1}{p}\right) = \begin{cases} 1, & \text{if } p \equiv 1 \pmod{4} \\ -1, & \text{if } p \equiv 3 \pmod{4}. \end{cases}$$

(ii) If $a \equiv b \pmod{p}$, *then* $\left(\frac{a}{p}\right) = \left(\frac{b}{p}\right)$.

(iii) (Multiplicative property) $\left(\frac{a \times b}{p}\right) = \left(\frac{a}{p}\right) \times \left(\frac{b}{p}\right)$.

(iv) $\left(\frac{a^2}{p}\right) = \begin{cases} 1, & \text{if } p \nmid a \\ 0, & \text{if } p \mid a \end{cases}$

(v) $\left(\frac{2}{p}\right) = (-1)^{(p^2-1)/8} = \begin{cases} 1, & \text{if } p \equiv 1 \text{ or } 7 \pmod 8 \\ -1, & \text{if } p \equiv 3 \text{ or } 5 \pmod 8). \end{cases}$

(vi) (Law of quadratic reciprocity) $\left(\frac{p}{q}\right) \times \left(\frac{q}{p}\right) = (-1)^{((p-1)/2)\times((q-1)/2)}$. *In other words,*

- *If* $p \equiv 1 \pmod 4$ *or* $q \equiv 1 \pmod 4$, *then* $\left(\frac{p}{q}\right) = \left(\frac{q}{p}\right)$.
- *If* $p \equiv q \equiv 3 \pmod 4$, *then* $\left(\frac{p}{q}\right) = -\left(\frac{q}{p}\right)$.

Note that a more useful form of the quadratic reciprocity law is

$$\left(\frac{p}{q}\right) = (-1)^{((p-1)/2)\times((q-1)/2)} \times \left(\frac{q}{p}\right).$$

If we assume that $p < q$ on the left-hand side, then q may be replaced by $q \pmod p$ on the right-hand side. It turns out that evaluating the right-hand side is a smaller problem. This procedure can be iterated until p and q get small.

Example 8

$$\left(\frac{37}{103}\right) = \left(\frac{103}{37}\right) = \left(\frac{29}{37}\right) = \left(\frac{37}{29}\right) = \left(\frac{8}{29}\right) = \left(\frac{2^2}{29}\right) \times \left(\frac{2}{29}\right) = 1 \times (-1) = -1.$$

Corollary 2 *Let p be an odd prime. The product of two quadratic residues modulo p is a residue, the product of two non-residues is a residue, and the product of a residue and a non-residue is a non-residue.*

The **Jacobi symbol** extends the domain of the denominator of the Legendre symbol, but its properties are not so simple. The Jacobi symbol, denoted by $\left(\frac{a}{n}\right)$, is a number-theoretic tool which is defined for all $a \geq 0$ and all odd positive integers n. Let $\prod_{i=1}^{r} p_i^{e_i}$ be the prime factorization of n. Then,

$$\left(\frac{a}{n}\right) = \prod_{i=1}^{r} \left(\frac{a}{p_i}\right)^{e_i}.$$

Here $\left(\frac{a}{p_i}\right)$ denotes the previously defined Legendre symbol. These two symbols coincide in the case where n is an odd prime. Like the Legendre symbol, the Jacobi symbol is equal to either 0, 1 or -1. If $\left(\frac{a}{n}\right) = -1$, then a is a non-residue modulo n, and if $\gcd(a,n) \neq 1$ then $\left(\frac{a}{n}\right) = 0$. However, if $\left(\frac{a}{n}\right) = 1$, unlike the Legendre symbol, a may be either a residue or a non-residue. For example, $\left(\frac{2}{15}\right) = 1$, but 2 is not a quadratic residue modulo 15. On the other hand, if a is a quadratic residue modulo n, then $\left(\frac{a}{n}\right) = 1$ is always satisfied. A quadratic non-residue a with $\left(\frac{a}{n}\right) = 1$ is called a pseudo-square. For deeper and further knowledge about the Jacobi and Legendre symbols and quadratic residues, one can read the related chapter of the book [23].

The following theorem, which can be easily proven by the Chinese Remainder Theorem, shows that knowledge of the factorization of n is sufficient to decide whether a given number in $\mathbb{Z}_n^*$ is a quadratic residue or not.

Theorem 18 *Let $n = p_1^{e_1} \times p_2^{e_2} \times \cdots \times p_r^{e_r} > 1$ be the prime factorization of n. Then, $a \in \mathbb{Z}_n^*$ is a quadratic residue modulo n if and only if a is a quadratic residue modulo p_i for all $i = 1, \ldots, r$.*

If $n > 3$ is an odd composite integer, the problem of determining whether a non-negative integer a with Jacobi symbol 1 is a quadratic residue modulo n is called the **Quadratic Residuosity Problem**, denoted by **QR[*n*]**. There is no polynomial-time procedure, without the factorization of n, to distinguish quadratic residues modulo n. Therefore, QR[n] can be polynomially reduced to Fact[n], i.e.

$$\text{Fact}\ [n] \Rightarrow\ \text{QR}\ [n].$$

However, we don't know whether the Integer Factorization Problem is polynomially reducible to the Quadratic Residuosity Problem.

Algorithm 8 Computation of the Jacobi symbol

```
1: function JAC(a, n)
2:     Input: Integer a ≥ 0, odd integer modulus n > 0.
3:     Output: The Jacobi symbol (a/n).
4:     j := 1
5:     if n is even then PRINT("n must be odd!")
6:     if a = 0 then j := 0
7:     while a ≠ 0 do
8:         while a is even do
9:             a := a/2
10:            if n ≡ 3 (mod 8) or n ≡ 5 (mod 8) then j := −j
11:        end while
12:        if a ≥ n then a := a (mod n)
13:        else a := n, n := a                                  ▷ Swap a and n.
14:            if a ≡ 3 (mod 4) and n ≡ 3 (mod 4) then j := −j
15:            a := a (mod n)
16:    end while
17:    if n = 1 then return j
18:    else return 0
```

2.1.11 Higher-Order Residues

An integer $a \in \mathbb{Z}_n^*$ is called an ***r*th residue** modulo n if there exists an integer $x \in \mathbb{Z}_n^*$ such that

$$a \equiv x^r \pmod{n}$$

and x is said to be an rth root of a. If there is no solution to this congruence, then a is called an ***r*th non-residue** modulo n.

Lemma 2 *[1, Lemma 2.1, Lemma 2.2]*

(i) The set of rth residues modulo n which are relatively prime to n is a subgroup of $\mathbb{Z}_n^*$.
(ii) Each rth residue modulo n in $\mathbb{Z}_n^*$ *has the same number of rth roots.*

Given a random $a \in \mathbb{Z}_n^*$, the problem of determining whether a is an rth residue modulo n is called the **Higher Residuosity Problem** (or ***r*th Residuosity Problem**). This problem is intractable when n is a composite integer whose factorization is unknown, since there is no known polynomial-time procedure to decide whether or not an integer is an rth residue modulo n.

The following lemma directly follows from Theorem 4 and Euler's Theorem (Theorem 11).

Lemma 3 *If* $\gcd(r, \phi(n)) = 1$, *then every integer* $y \in \mathbb{Z}_n^*$ *is an rth residue modulo n.*

Paillier [19] considered the Higher Residuosity Problem in the special case when r is replaced by n and n is replaced by n^2 for $n = p \times q$, where p and q are distinct odd primes. This special case of the problem is called the **Composite Residuosity Problem**, denoted by $\mathbf{CR}[n]$, which is intractable.

2.1.12 Residue Classes

Let (r, n, y) be fixed integers with $y \in \mathbb{Z}_n^*$. If $w \in \mathbb{Z}_n^*$ is expressible as $w \equiv y^m \times u^r \pmod{n}$ for some $u \in \mathbb{Z}_n^*$, then w is said to be of **residue class** m (with respect to y) or m is the ***r*th residue class of** w (with respect to y). The set of all elements in $\mathbb{Z}_n^*$ which are of residue class m is denoted by

$$RC[m] = \{w \in \mathbb{Z}_n^* : w \equiv y^m \times u^r \pmod{n} \text{ for some } u \in \mathbb{Z}_n^*\}.$$

In particular, $RC[0]$ is the set of rth residues modulo n.

The following lemma shows that for a given (r, n, y), any two residue classes either coincide or are disjoint.

Lemma 4 *[1, Lemma 2.4] Let* (r, n, y) *be fixed integers with* $y \in \mathbb{Z}_n^*$. *If* $RC[m_1] \cap RC[m_2]$ *is not empty, then* $RC[m_1] = RC[m_2]$.

The next lemma gives some important properties of residue classes, but its proof is simple and directly follows from the definition of residue class. To see its proof, see Lemma 2.8 and Lemma 2.9 in [1] for part (i) and part (ii), respectively.

Lemma 5 *Let (r,n,y) be fixed integers with $y \in \mathbb{Z}_n^*$.*

(i) If $w_1 \in RC[m_1]$ and $w_2 \in RC[m_2]$, then $w_1 \times w_2 \in RC[m_1 + m_2]$.
(ii) If $w \in RC[m]$, then $w^{-1} \in RC[-m]$.

The following lemma follows from the above lemma and shows how two integers can be of the same residue class.

Lemma 6 *[1, Lemma 2.10] Let (r,n,y) be fixed integers with $y \in \mathbb{Z}_n^*$. Two integers w_1 and w_2 are of the same residue class if and only if $w_1 \times {w_2}^{-1}$ is an rth residue modulo n.*

Lemma 7 *[1, Lemma 2.5, 2.6] Let (r,n,y) be integers with $y \in \mathbb{Z}_n^*$. Let k be the least positive integer such that y^k is an rth residue modulo n. Then the following hold:*

(i) k divides r.
(ii) $RC[m_1] = RC[m_2]$ if and only if $m_1 \equiv m_2 \pmod{k}$.
(iii) There are exactly k distinct residue classes and they are given by

$$RC[0], RC[1], \ldots, RC[k-1].$$

We now present some necessary conditions for r to be the least positive integer making y^r an rth residue modulo n. See Lemma 2.12, Corollary 2.14 and Theorem 2.17 in [1] for the proofs of part (i), part (ii), and part(iii) of the following result, respectively.

Lemma 8 *Let (r,n,y) be integers with $y \in \mathbb{Z}_n^*$. If r is the least positive integer such that y^r is an rth residue modulo n, then the following hold:*

(i) y^s is an rth residue modulo n if and only if $s \equiv 0 \pmod{r}$.
(ii) $r \mid \phi(n)$.
(iii) Each $w \in \mathbb{Z}_n^*$ is expressible as $w \equiv y^m \times u^r \pmod{n}$ for at most one integer m with $0 \le m < r$ and at most one $u \in \mathbb{Z}_n^*$. Hence, $RC[0], RC[1], \ldots, RC[r-1]$ are all distinct and represent the entire set of residue classes.

We know that each rth residue modulo n has the same number of rth roots (see Lemma 2.2 in [1]). The following theorem says more.

Theorem 19 *[1, Theorem 2.20] Let $y \in \mathbb{Z}_n^*$ be an rth residue modulo n. If $r \mid \phi(n)$ and $\gcd(r, \phi(n)/r) = 1$, then y has exactly r distinct rth roots.*

Since $|\mathbb{Z}_n^*| = \phi(n)$, the following two results follow immediately.

Corollary 3 *If $r \mid \phi(n)$ and $\gcd(r, \phi(n)/r) = 1$, then the number of rth residues modulo n is $\phi(n)/r$.*

Corollary 4 *If $r \mid \phi(n)$ and $\gcd(r, \phi(n)/r) = 1$, then $y \in \mathbb{Z}_n^*$ is an rth residue modulo n if and only if $y^{\phi(n)/r} \equiv 1 \pmod{n}$.*

If we put the further assumption of $\gcd(r, \phi(n)/r) = 1$ into the assumption of Lemma 8, then $RC[0], RC[1], \ldots, RC[r-1]$ partition $\mathbb{Z}_n^*$. This follows from the following theorem.

Theorem 20 *[1, Theorem 2.22] Let (r, n, y) be integers with $y \in \mathbb{Z}_n^*$. If r is the least positive integer such that y^r is an rth residue modulo n and $\gcd(r, \phi(n)/r) = 1$, then every $w \in \mathbb{Z}_n^*$ can be expressed as $w \equiv y^m \times u^r \pmod{n}$ for a unique integer $m \in \mathbb{Z}_r$ and a unique $u \in \mathbb{Z}_n^*$, i.e. every $w \in \mathbb{Z}_n^*$ is a member of exactly one residue class $RC[m]$ for some $0 \leq m < r$.*

If the unique rth residue class of w is m for given integers r, n and $y \in \mathbb{Z}_n^*$, then $[\![w]\!]_y$ denotes this unique class m, where

$$[\![\ ]\!]_y : (\mathbb{Z}_n^*, \times) \to (\mathbb{Z}_n, +)$$
$$w \mapsto [\![w]\!]_y$$

is the class function.

The problem of determining to which rth residue class w belongs is conjectured to be a difficult problem. This is called **the rth Residuosity Class Problem**. Paillier [19] considered this problem in the special case when r is replaced by n and n is replaced by n^2 for $n = p \times q$, where p and q are distinct odd primes. Given $g, w \in \mathbb{Z}_n^*$ such that $n \mid \mathrm{ord}(g)$, computing the class $[\![w]\!]_g$ is called the **Composite Residuosity Class Problem** of base g, denoted by **Class $[n, g]$**, on which Paillier's algorithm is based. Since all the instances of Class$[n, g]$ are computationally equivalent for every g by Lemma 7 in [19], the complexity of Class $[n, g]$ is independent of g. So it is simply denoted by **Class $[n]$**. The computational problem Class $[n]$ is intractable due to the nonexistence of a probabilistic polynomial-time algorithm solving it. Paillier described an important variant of Class $[n, g]$, which is the **Decisional Composite Residuosity Class Problem**, denoted by **D-Class $[n]$**: Given $w \in \mathbb{Z}_{n^2}^*$, $g \in \mathbb{Z}_n^*$ such that $n \mid \mathrm{ord}(g)$ and $x \in \mathbb{Z}_n$, decide whether $x = [\![w]\!]_y$ or not.

A hierarchy of number-theoretic problems is demonstrated by Paillier [19] in Theorem 9, Theorem 11 and Theorem 12 as follows:

$$\text{CR}\,[n] \equiv \text{D-Class}\,[n] \Leftarrow \text{Class}\,[n] \Leftarrow \text{RSA}\,[n, n] \Leftarrow \text{Fact}\,[n].$$

The equivalence of CR$[n]$ and D-Class $[n]$ follows from the following fact.

Proposition 2 $[\![w]\!]_g = 0$ *if and only if w is an nth residue modulo n^2.*

Proof. Let w be an nth residue modulo n^2. We can write w as

$$w \equiv g^{[\![w]\!]_g} \times u^n \pmod{n^2}$$

for some $u \in \mathbb{Z}_n^*$. Note that w is an nth residue modulo n^2 if and only if $g^{[\![w]\!]_g}$ is an nth residue. Since $[\![w]\!]_g \in \mathbb{Z}_n$, the class $[\![w]\!]_g$ should be less than n, which means $g^{[\![w]\!]_g}$ cannot be an nth residue modulo n^2 unless $[\![w]\!]_g = 0$.

Conversely, if $[\![w]\!]_g = 0$, then there exists $u \in \mathbb{Z}_n^*$ such that

$$w \equiv g^0 \times u^n \equiv u^n \pmod{n^2},$$

which means w is an nth residue modulo n^2.

2.1.13 Random Number Generators

In the context of homomorphic encryption functions, random positive integers in particular ranges are required. We define the function RANDINT(a,b) that returns a uniformly selected integer in the range $[a,b]$. How does one accomplish this task?

It is intuitively clear that random numbers should be unpredictable, even for an adversary who has collected a large set of random numbers previously generated by the same source. The RANDINT function requires that random numbers should be uniformly distributed on their range and independent. This characterizes an ideal random number generator, which is a mathematical construction. In practice, we require that (i) the random numbers should have good statistical properties, (ii) the knowledge of subsequences of random numbers shall not allow one to practically guess predecessors or successors [10].

Real-world random number generators fall into two classes: deterministic random number generators (DRNGs) and true random number generators (TRNGs). The latter class is based on specially prepared physical processes, such as nondeterministic effects of electronic circuits, for example, noise from a Zener diode, inherent semiconductor thermal noise, free-running oscillators, etc. Meanwhile, DRNGs are algorithmic methods, that is, they compute numbers. TRNGs are safer against guessing attacks, as they rely on the unpredictability of their output (theoretical security) while the security of DRNGs essentially depends on the computational complexity of possible attacks (practical security). There are also hybrid RNGs based on methods from both fields.

For the purpose of implementing a RANDINT function to be used in encryption functions in this book, a pure DRNG can be used. That starts with an internal state initially equal to a seed s_0 which is possibly generated by a TRNG [26], and updates the state n times using $s_i = \Phi(s_{i-1})$ for $i = 1, 2, \ldots, n$ where $\Phi(\cdot)$ is the state transition function. Random numbers r_i for $i = n, n+1, \ldots$ are generated using both the state transition function and the output transition function $\Psi(\cdot)$ as

$$\begin{aligned} r_i &= \Psi(s_i) \\ s_{i+1} &= \Phi(s_i). \end{aligned}$$

Choices for $\Phi(\cdot)$ and $\Psi(\cdot)$ include block ciphers and hash functions, satisfying security assumptions, statistical randomness and independence properties.

2.2 Group Theory

In this section, a basic algebraic structure, group, is introduced. Some examples, properties, and related mathematical problems are presented.

2.2.1 Basic Axioms

Let G be a set. A **binary operation** on G is a function

$$\begin{aligned} *: G \times G &\to G \\ (a,b) &\mapsto a * b. \end{aligned}$$

We say $*$ is **associative** if for all $a, b \in G$, we have $a * (b * c) = (a * b) * c$, and $*$ is **commutative** if for all $a, b \in G$, $a * b = b * a$. For example, $+$ (usual addition) is a commutative binary operation on $\mathbb{Z}$, and $\times$ (usual multiplication) is a commutative binary operation on $\mathbb{Z}$. If the restriction of $*$ to a subset H of G is a binary operation on H, i.e. $a * b \in H$ for all $a, b \in H$, then H is said to be **closed** under $*$.

A **group** is an ordered pair $(G, *)$, where G is a nonempty set and $*$ is a binary operation, satisfying the following axioms:

(i) G is closed under $*$, i.e. $a * b \in G$ for all $a, b \in G$.
(ii) $*$ is associative, i.e. $a * (b * c) = (a * b) * c$ for all $a, b \in G$.
(iii) there exits a unique element e in G, called the **identity** of G, such that $a * e = e * a = a$ for all $a \in G$.
(iv) for each $a \in G$ there is a uniquely determined element $a^{-1} \in G$, called the **inverse** of a, such that $a * a^{-1} = a^{-1} * a = e$.

It is easy to verify that $(a^{-1})^{-1} = a$ for all $a \in G$, and $(a * b)^{-1} = (b^{-1}) * (a^{-1})$.

The group $(G, *)$ is called **commutative** (or **abelian**) if $a * b = b * a$ for all $a, b \in G$. If G is a finite set, then $(G, *)$ is a finite group.

2.2.2 Multiplicative Groups

The group operation on G is $\times$ (usual multiplication). The identity element is generally called the **unit element**, which is 1. Multiplication of an element a by itself k times is denoted by $a^k = a \times a \times \cdots \times a$ (k copies). Let $a^0 = 1$, the identity of G. The inverse of an element a is denoted by a^{-1}.

Example 9 *For any integer $n > 0$, $(\mathbb{Z}_n^*, \times)$ is a multiplicative group, where $\mathbb{Z}_n^*$ is the set of invertible elements modulo n and $\times$ is the multiplication modulo n. Consider the multiplication tables for modulo 5 and modulo 6:*

$\times$ (mod 5)	1	2	3	4
1	**1**	2	3	4
2	2	4	**1**	3
3	3	**1**	4	2
4	4	3	2	**1**

$\times$ (mod 6)	1	2	3	4	5
1	**1**	2	3	4	5
2	2	4	0	2	4
3	3	0	3	0	3
4	4	2	0	4	2
5	5	4	3	2	**1**

While multiplication modulo 5 on the set $\mathbb{Z}_5 - \{0\} = \{1,2,3,4\}$ *forms the group* $(\mathbb{Z}_5^*, \times)$*, multiplication modulo 6 on the set* $\mathbb{Z}_6 - \{0\} = \{1,2,3,4,5\}$ *does not form a group since 2, 3 and 4 are not invertible. However, multiplication modulo 6 on the set of invertible elements forms a group:* $(\mathbb{Z}_6^*, \times) = (\{1,5\}, \times)$.

2.2.3 Additive Groups

The group operation on G is $+$ (usual addition). The identity element is generally called the **zero element**, which is 0. Addition of an element a to itself k times is denoted by $k \times a = a + \cdots + a$ (k copies). Let $0 \times a = 0$, the identity of G. The inverse of an element a is denoted by $-a$.

Example 10 *For any integer* $n > 0$, $(\mathbb{Z}_n, +)$ *is an additive group, where* $+$ *is the addition modulo n. Consider the addition tables for modulo 4 and modulo 5:*

$+$ (mod 4)	0	1	2	3
0	0	1	2	3
1	1	2	3	0
2	2	3	0	1
3	3	0	1	2

$+$ (mod 5)	0	1	2	3	4
0	0	1	2	3	4
1	1	2	3	4	0
2	2	3	4	0	1
3	3	4	0	1	2
4	4	0	1	2	3

Addition modulo 4 on the set $\mathbb{Z}_4 = \{0,1,2,3\}$ *forms the group* $(\mathbb{Z}_4, +)$ *and addition modulo 5 on the set* $\mathbb{Z}_5 = \{0,1,2,3,4\}$ *forms the group* $(\mathbb{Z}_5, +)$.

2.2.4 Order of a Group

The **order of a group** $(G, *)$ is the number of elements in the set G, denoted by $\mathrm{ord}(G)$ or $|G|$, which is either a positive integer or ∞.

Example 11 *For a prime number p, the order of* $(\mathbb{Z}_p^*, \times)$ *is equal to* $p-1$. *For any composite integer* $n > 0$, $\mathrm{ord}(\mathbb{Z}_n^*) = \phi(n)$. *The order of* $(\mathbb{Z}_n, +)$ *is equal to n, where n is either prime or composite.*

The **order of an element** a in a multiplicative group is the smallest integer k such that $a^k = 1$, where 1 is the unit element of the group. For example, $\text{ord}(3) = 5$ in $(\mathbb{Z}_{11}^*, \times)$ since $\{3^i \pmod{11} \mid 1 \leq i \leq 10\} = \{3,9,5,4,1\}$. Note that $\text{ord}(1) = 1$.

Theorem 21 *Let $g \in G$ be of finite order, where G is a multiplicative group, and $m, k \in \mathbb{Z}$. Then,*

(i) $g^k = 1$ *if and only if* $\text{ord}(g)$ *divides* k.
(ii) $g^k = g^m$ *if and only if* $k \equiv m \pmod{\text{ord}(g)}$.
(iii) $\text{ord}(g^m) = \text{ord}(g)/\gcd(\text{ord}(g), m)$.

The **order of an element** a in an additive group is the smallest integer k such that $k \times a = a + \cdots + a = 0$, where 0 is the identity element. For example, $\text{ord}(3) = 11$ in $(\mathbb{Z}_{11}, +)$ since $\{i \times 3 \pmod{11} \mid 1 \leq i \leq 11\} = \{3,9,6,1,4,7,10,2,5,8,0\}$. Note that $\text{ord}(0) = 1$.

An element whose order is equal to the group order is called **primitive**. For any positive integer n, the primitive elements of $(\mathbb{Z}_n, +)$ are precisely those x with $\gcd(x, n) = 1$. In particular, if n is prime, then all elements in $(\mathbb{Z}_n, +)$ are primitive. Since $|\mathbb{Z}_n^*| = \phi(n)$, from Theorem 21, part (iii), we can conclude that if a is a primitive element of $(\mathbb{Z}_n^*, \times)$, then a^i is also a primitive element if and only if $\gcd(i, \phi(n)) = 1$.

Example 12 *All the elements in $(\mathbb{Z}_{11}, +)$ are primitive as 11 is prime. The order of the group $(\mathbb{Z}_{11}^*, \times)$ is 10 and* $\text{ord}(2) = 10$, *hence 2 is a primitive element of this multiplicative group.*

2.2.5 Cyclic Groups and Subgroups

A nonempty subset H of G is called a **subgroup** of G if H together with the group operation of G is a group. For finite groups, it suffices to check for closure under the operation and taking of inverses. Both conditions can be checked at the same time via one expression, as H is a subgroup of G if and only if $x * y^{-1} \in H$ for all $x, y \in H$.

For any $g \in G$, let g^k denote the repeated application of the group operation on g, namely $g * g \cdots * g$, k times. The set $\{g^k \mid k \in \mathbb{Z}\}$ is a subgroup of G. It is called the **subgroup generated** by g and is denoted by $\langle g \rangle$. If g has a finite order, then $\langle g \rangle = \{g^k \mid 0 \leq k < \text{ord}(g)\}$. If $G = \langle g \rangle$ for some $g \in G$, then G is called **cyclic** and g is called a **generator** of G. Any primitive element of G is a generator of G. If G is an additive group, we use $k \times g$ instead of g^k.

Example 13 *2 is a generator of $(\mathbb{Z}_{11}^*, \times)$ since*

$$\langle 2 \rangle = \{2^k \mid 1 \leq k \leq 10\} = \{2,4,8,5,10,9,7,3,6,1\} = \mathbb{Z}_{11}^*.$$

Also, 2 is a generator of $(\mathbb{Z}_{11},+)$ *since*

$$\langle 2\rangle = \{k\times 2 \mid 1\le k\le 11\} = \{2,4,6,8,10,1,3,5,7,9,0\} = \mathbb{Z}_{11}.$$

Theorem 22 *If G is finite and cyclic, then G has exactly* $\phi(|G|)$ *generators.*

Example 14 *There are* $\phi(10) = 4$ *generators of the group* $(\mathbb{Z}_{11}^*,\times)$*, which are* $2,6,7,8$ *all of which are of order 10.*

We next record all the subgroups of $\mathbb{Z}$.

Theorem 23 *Any subgroup of* $\mathbb{Z}$*, where* $(\mathbb{Z},+)$ *is the additive group of integers, is of the form*

$$\langle n\rangle = n\mathbb{Z} = \{\dots,-2\times n,-n,0,n,2\times n,\dots\}$$

for a unique integer $n\ge 0$*. Thus all subgroups of* $\mathbb{Z}$ *are cyclic.*

2.2.6 Discrete Logarithm Problem

Let G be a finite cyclic group of order n and g be a generator of this group. Given a group element $\alpha = g^x \in G$, the **discrete logarithm** of α to the base g, denoted by $\log_g \alpha$, is the integer $0\le x<n$. Given α, the problem of finding this integer x is called the **Discrete Logarithm Problem (DLP)**.

The DLP is believed (although not proven) to be intractable. The hardness of the DLP is independent of the choice of the generator. However, the hardness of the DLP depends on the groups chosen. The DLP is trivial in the additive modulo p group, $(\mathbb{Z}_p,+)$, due to the Extended Euclidean Algorithm. The multiplicative cyclic group $\mathbb{Z}_p^*$, where p is a prime, is a popular choice for groups used in cryptosystems whose security is based on the DLP. If the order of the group has small prime factors, namely if $\phi(p) = p-1 = {q_1}^{a_1}\times\cdots\times {q_k}^{a_k}$, where all the q_i are small primes, then DLP can be solved in this group easily by the Pohlig-Hellman algorithm (see [14], Sect. 3.6.4) via quickly computing $x \pmod{{q_i}^{a_i}}$ by the Chinese Remainder Theorem. So p must be chosen such that $p-1$ has at least one large prime factor q. This guarantees that the Pohlig-Hellman algorithm cannot solve the DLP easily. A prime p of the form $2\times q+1$, where q is a large prime, is called a safe prime. Even if p is a safe prime, there is a sub-exponential algorithm solving DLP which is called the index calculus (see [14], Sect. 3.6.5). It turns out that for the safety of the cryptosystem, p must be a very large prime (at least 1024 bits).

2.2.7 Diffie-Hellman Problem

Let G be a finite cyclic group of order p and g be a generator of this group. Given two group elements g^a and g^b, the problem of computing $g^{a\times b}$ without knowledge of a and b is called the **Diffie-Hellman Problem (DHP)**. To be able to compute $g^{a\times b}$, either a or b must be given. Solving for either exponent is the DLP, which is intractable for carefully chosen groups. Sometimes the DHP is called the Computational Diffie-Hellman Problem (CDHP). The security of the Diffie-Hellman key exchange protocol is based on the difficulty of solving DHP in $\mathbb{Z}_p^*$.

The most significant variant of DHP is the **Decisional Diffie-Hellman Problem (DDHP)**. Consider the group G above. The problem is, given three group elements g^a, g^b and g^c, to determine whether $g^{a\times b} = g^c$ without knowledge of a and b. In other words, the DDHP is the problem of determining whether (g^a, g^b, g^c) is a D-H triplet or not. The security of the ElGamal cryptosystem is based on both CDHP and DDHP.

There is a relation between DLP, CDHP and DDHP. Let us be given an oracle solving DLP. Then y is computed from g^y and then $g^{x\times y}$ is computed by raising g^x to the power y, therefore CDHP is solved. In other words, CDHP is not harder than DLP. Since it is easier to verify a solution than to compute it, solving CDHP directly enables solving DDHP. So, DDHP is not harder than CDHP. It turns out that we have the following hierarchy between these problems:

$$\text{DLP} \Rightarrow \text{CDHP} \Rightarrow \text{DDHP}\,.$$

2.2.8 *Cosets and Lagrange's Theorem*

A **left coset** of a subgroup H in a group G with representative $g \in G$ is the set $gH = \{g * h \mid h \in H\}$. Two left cosets are either disjoint or equal. So the set of left cosets partitions G. Furthermore, $gH = g'H$ if and only if $g^{-1} * g' \in H$. In particular, $gH = H$ if and only if $g \in H$. The set of left cosets of H in G forms the **quotient group** or the **factor group**, denoted by G/H, with operation defined by $(gH) \star (g'H) = (g * g')H$. This operation is well-defined so that the operation does not depend on the choice of representatives for the cosets. Consider the bijection

$$\begin{aligned} H &\to gH \\ h &\mapsto g * h, \end{aligned}$$

which yields the fact that all left cosets have the same cardinality, which is $|H|$. The number of left cosets is denoted by $[G:H] = |G/H|$, which is called the **index** of H in G. If G is finite, then G is partitioned into $[G:H]$ cosets, each of cardinality $|H|$. It follows that $[G:H] = |G|/|H|$. This can be derived as a consequence of a more general result which is known as Lagrange's Theorem.

Theorem 24 ***(Lagrange's Theorem)*** *If H is a subgroup of a finite group G, then $|H|$ divides $|G|$.*

A **right coset** of a subgroup H in a group G with representative $g \in G$ is the set $Hg = \{h * g \mid h \in H\}$. All the statements above for left cosets are true with "right coset" in place of "left coset". If G is an abelian group, left cosets and right cosets of H in G are equal.

Since the order of an element is the order of the cyclic group it generates, we have the following.

Corollary 5 *For any* $a \in G$, $\mathrm{ord}(a) \mid \mathrm{ord}(G)$.

Example 15 *Consider the additive group of integers* $\mathbb{Z}$ *and let us consider the subgroup* $\langle n \rangle = n\mathbb{Z}$. *If* $k \in \mathbb{Z}$, *then* $k = q \times n + r$ *for integers* q, r *with* $0 \leq r < n$. *Since* $q \times n \in n\mathbb{Z}$, *then* $k + n\mathbb{Z} = r + q \times n + n\mathbb{Z} = r + n\mathbb{Z}$. *It turns out that any coset* $k + n\mathbb{Z}$ *is the same as the coset* $r + n\mathbb{Z}$, *where* r *is the remainder when* k *is divided by* n. *If* $0 \leq r_1 < r_2 < n$, *then* $0 < r_2 - r_1 < n$ *and so* $r_2 - r_1 \notin n\mathbb{Z}$, *i.e.* $r_1 + n\mathbb{Z} \neq r_2 + n\mathbb{Z}$. *Hence we see that* $\mathbb{Z}/n\mathbb{Z} = \{0 + n\mathbb{Z}, 1 + n\mathbb{Z}, \ldots, (n-1) + n\mathbb{Z}\}$, *where each element is just a congruence class modulo* n. *Therefore,* $\mathbb{Z}/n\mathbb{Z} = \mathbb{Z}_n$. *Note that the index of* $n\mathbb{Z}$ *in* $\mathbb{Z}$ *is* $n = |\mathbb{Z}_n|$.

2.2.9 Sylow's Theorem

Let G be a finite group and p be a prime number. A group of order p^a for some $a \geq 0$ is called a p-**group**. Subgroups of G which are p-groups are called p-**subgroups**.

The next theorem is known as Sylow's Theorem.

Theorem 25 *Let* G *be a finite group of order* $p^k \times m$, *where* $p \nmid m$, *then there exists at least one subgroup of order* p^k. *Such subgroups are called* **Sylow** p-**subgroups**.

Corollary 6 *If* G *is an abelian (commutative) group, then there is a unique Sylow* p-*subgroup of* G.

Example 16 *There is a unique Sylow* 7-*subgroup of* $\mathbb{Z}_{49}^*$ *as the multiplicative group* $\mathbb{Z}_{49}^*$ *is abelian and* $\left|\mathbb{Z}_{49}^*\right| = 42 = 6 \times 7$. *The Sylow* 7-*subgroup of* $\mathbb{Z}_{49}^*$ *is of order* 7 *and it is*

$$H_7 = \langle 8 \rangle = \{1, 8, 15, 22, 29, 36, 43\}.$$

For any group $\mathbb{Z}_n^*$ with $n = p^2 \times q$, where p and q are distinct primes, there is a unique p-subgroup. Because $|\mathbb{Z}_n^*| = q \times p \times (p-1)$ among powers of p the only possible divisor of $|\mathbb{Z}_n^*|$ is p. The problem of determining whether an element g of $\mathbb{Z}_n^*$ belongs to the unique p-subgroup of $\mathbb{Z}_n^*$ is called **the** p-**Subgroup Problem**, which is denoted by **PSUB**.

2.2.10 Direct Products

The **direct product** $G_1 \times G_2 \times \cdots \times G_n$ of the groups $G_1, G_2, \ldots, G_n$ with operations $*_1, *_2, \ldots, *_n$, respectively, is the set of n-tuples $(g_1, g_2, \ldots, g_n)$ where $g_i \in G_i$ with operation defined componentwise:

$$(g_1, g_2, \ldots, g_n) * (h_1, h_2, \ldots, h_n) = (g_1 *_1 h_1, g_2 *_2 h_2, \ldots, g_n *_n h_n).$$

Theorem 26 *If $G_1, G_2, \ldots, G_n$ are groups, then their direct product $G_1 \times G_2 \times \cdots \times G_n$ is a group of order $|G_1| \times |G_2| \times \cdots \times |G_n|$.*

2.2.11 Homomorphisms

Given two groups $(G, *)$ and $(H, \diamond)$, a **group homomorphism** from $(G, *)$ to $(H, \diamond)$ is a map $f : (G, *) \to (H, \diamond)$ such that

$$f(x * y) = f(x) \diamond f(y),$$

for all $x, y \in G$. In other words, a map is a homomorphism if it respects the group structures of its domain and codomain. In addition to the homomorphic property, if f is also a bijection, then f is called a **group isomorphism** and G and H are called isomorphic, written as $G \cong H$.

Example 17 *Consider the exponential map* $\exp : (\mathbb{R}, +) \to (\mathbb{R}^+, \times)$ *from the additive group of real numbers to the multiplicative group of nonzero real numbers defined by* $\exp(x) = e^x$*, where e is the base of the natural logarithm.* exp *is a homomorphism since* $\exp(x+y) = e^{x+y} = e^x \times e^y = \exp(x) \times \exp(y)$.

Group homomorphisms are the basis of partially homomorphic cryptosystems. Let (P, C, K, E, D) be a cryptosystem (or an encryption scheme), where P is the plaintext or message space, C is the ciphertext space, K is the key space, E is the encryption function and D is the decryption function (or algorithm). Assume that the plaintexts form a group under $*$, and the ciphertexts form a group under $\diamond$. Then the encryption function is

$$E : (P, *) \to (C, \diamond).$$

If $D(E(m_1 * m_2)) = D(E(m_1) \diamond E(m_2))$, then the encryption scheme is homomorphic.

A cryptosystem that has the encryption function E is called **additively homomorphic** if there exists an operation $\boxplus$ such that

$$D(E(m_1 + m_2)) = D(E(m_1) \boxplus E(m_2)), \tag{2.1}$$

and it is called **multiplicatively homomorphic** if there exists an operation $\boxtimes$ such that

$$D(E(m_1 \times m_2)) = D(E(m_1) \boxtimes E(m_2)). \tag{2.2}$$

Indeed, an encryption scheme that has the homomorphic property supporting only one type of operations, either addition or multiplication, is called a **partially homomorphic encryption scheme**. If a cryptosystem supports both (2.1) and (2.2) it is called a **fully homomorphic encryption scheme**. If a homomorphic encryption scheme satisfies only one of (2.1) and (2.2) infinitely many times while it supports the other a limited number of times, then it is called a **somewhat homomorphic encryption scheme**. A homomorphic encryption scheme may also preserve scalar multiplication by an integer scalar k such that

$$\text{either } D(E(k \times m)) = D(E(m)^k) \text{ or } D(E(k \times m)) = D(k \times E(m)),$$

or exponentiation by an integer scalar k such that

$$\text{either } D(E(m^k)) = D(E(m)^k) \text{ or } D(E(m^k)) = D(k \times E(m)),$$

or addition with an integer scalar k such that

$$D(E(k+m)) = D(E(m) \times \alpha^k)$$

where α is a group element.

2.3 Field Theory

A **field** is a nonempty set F together with two commutative binary operations $+$ and $\times$ such that

(i) $(F,+)$ is an abelian group with identity 0,
(ii) $(F-\{0\},\times)$ is an abelian group with identity 1,
(iii) the following distributive law holds: $a \times (b+c) = (a \times b) + (a \times c)$, for all $a,b,c \in F$.

For any field F, let $\mathbf{F}^* = \mathbf{F} - \{\mathbf{0}\}$. If the number of elements in F, called the **size** of F and denoted by $|F|$, is finite then F is called a finite field; otherwise, it is called an infinite field. If there exists a positive integer p such that $p \times 1 = 0$ (1 and 0 are the identity elements of the groups $(F^*,\times)$ and $(F,+)$, respectively) and p is the smallest such integer, then p is called the **characteristic** of F and denoted by $\text{char}(F) = p$. Note that the smallest such integer is necessarily prime. It turns out that if $k \times 1 = 0$ then k is divisible by p. If no such integer exists, then $\text{char}(F) = 0$.

Example 18 *The set of real numbers $\mathbb{R}$ and the set of complex numbers $\mathbb{C}$ are infinite fields with zero characteristic. However, the set of integers $\mathbb{Z}$ with the integer* *addition and multiplication operations does not form a field due to the fact that* $(\mathbb{Z}^*, \times)$ *is not a multiplicative group since none of the integers except 1 and -1 has a multiplicative inverse in* $\mathbb{Z}$.

Lemma 9 *If* $\text{char}(F) = p$, *then* $|F| = p^n$ *for some integer* $n = 1, 2, 3, \ldots$

Example 19 *The set of congruence classes modulo a prime* p, *namely* $\mathbb{Z}_p$, *is a finite field with both size and characteristic* p. *This field is called a* ***finite field with p elements*** *and denoted also by* $\mathbf{GF(p)}$ *or* $\mathbb{F}_p$.

If K is a nonempty subset of a field F and K is itself a field, then K is called a **subfield** of F. Here, F is called an **extension field** of K and denoted by F/K. The **degree** (or **index**) of a field extension F/K, denoted by $[F : K]$, is the dimension of F as a vector space over K. The extension is said to be finite if $[K : F]$ is finite and said to be infinite otherwise.

For a field extension F/K, the element $\alpha \in F$ is said to be **algebraic** over K if α is a root of some nonzero polynomial $p(x)$ whose coefficients are in K. If every element in F is algebraic over K, then the extension F/K is called an **algebraic extension.**

Lemma 10 *If the extension* F/K *is finite, then it is algebraic.*

The field $\overline{F}$ is called an **algebraic closure** of F provided that $\overline{F}$ is algebraic over F and every polynomial $p(x)$ with coefficients in F has a root in $\overline{F}$. A field F is **algebraically closed** if every polynomial with coefficients in F has a root in F.

Lemma 11 *Let* $\overline{F}$ *be an algebraic closure of* F. *Then* $\overline{F}$ *is algebraically closed.*

Lemma 12 *For any field* F *there exists an algebraically closed field containing* F.

Most popular finite fields that are used in cryptographic applications based on elliptic-curve-based schemes are prime fields $\text{GF}(p)$, binary extension fields $\text{GF}(2^n)$ and ternary extension fields $\text{GF}(3^n)$. The elements of the binary extension field $\text{GF}(2^n)$ can be represented as binary polynomials of degree at most $n-1$, whose coefficients are from $\text{GF}(2) = \{0, 1\}$. The addition in $\text{GF}(2^n)$ is modulo 2 addition of the corresponding coefficients of two polynomials. The multiplication in $\text{GF}(2^n)$ is polynomial multiplication followed by reduction by an irreducible binary polynomial of degree n. Similarly, the elements of the ternary extension field $\text{GF}(3^n)$ can be represented as ternary polynomials of degree at most $n-1$ whose coefficients are from $\text{GF}(3) = \{0, 1, 2\}$. The addition in $\text{GF}(3^n)$ is modulo 3 addition of the corresponding coefficients of two polynomials. The multiplication in $\text{GF}(3^n)$ is polynomial multiplication followed by reduction by an irreducible ternary polynomial of degree n.

2.4 Elliptic Curves

An **elliptic curve** $\mathscr{E}$ over a field K is the set of solutions of a cubic polynomial equation in two unknowns

$$\mathcal{E} : y^2 + a_1xy + a_3y = x^3 + a_2x^2 + a_4x + a_6, \tag{2.3}$$

where $a_1, a_2, a_3, a_4, a_6 \in K$ and $\Delta \neq 0$. Here Δ is the **discriminant** of $\mathcal{E}$ and is defined with the following set of equations:

$$\begin{aligned}
d_2 &= a_1^2 + 4 \times a_2, \\
d_4 &= 2 \times a_4 + a_1 \times a_3, \\
d_6 &= a_3^2 + 4 \times a_6, \\
d_8 &= a_1^2 \times a_6 + 4 \times a_2 \times a_6 - a_1 \times a_3 \times a_4 + a_2 \times a_3^2 - a_4^2, \\
\Delta &= -d_2^2 \times d_8 - 8 \times d_4^3 - 27 \times d_6^2 + 9 \times d_2 \times d_4 \times d_6.
\end{aligned}$$

Equation (2.3) is called a **long Weierstrass** equation. The field in which such an equation is solved can be a finite field or an infinite field, such as $\mathbb{C}$ (complex numbers), $\mathbb{R}$ (real numbers) or $\mathbb{Q}$ (rational numbers). The Weierstrass equations are also called Weierstrass curves. To ensure the smoothness of a Weierstrass curve, we require the discriminant Δ to be nonzero. If $\Delta = 0$, then the curve is singular, i.e. geometrically speaking, the curve has self-intersections and cusps.

It is possible to transform the long Weierstrass equation into the **short Weierstrass** equation

$$y^2 = x^3 + ax + b$$

by using some change of variables. The smoothness of a short Weierstrass curve is ensured by $\Delta = 4 \times a^3 + 27 \times b^2 \neq 0$.

Since $\lim_{x\to\infty} y = \infty$, the **point at infinity**, denoted by $\mathcal{O} = (\infty, \infty)$, is also considered to be a solution to the Weierstrass equation. If we want to consider points with coordinates in some extension field $K \supseteq F$, we write $\mathcal{E}(K)$, where

$$\mathcal{E}(K) = \{(x,y) \in K \times K \mid y^2 = x^3 + ax + b\} \cup \{\mathcal{O}\}.$$

2.4.1 Group Law

First of all, in order to define the addition operation $\oplus$ on an elliptic curve, we need the following simplified version of Bézout's Theorem.

Theorem 27 ***(Bézout's Theorem)*** *A line that intersects an elliptic curve at two points crosses at a third point.*

There is a **chord-and-tangent rule** for adding two points on a Weierstrass curve to get a third point on the curve. This addition operation enables us to find points on an elliptic curve when we are given only one or two.

The **point addition** is a geometric operation by Bézout's Theorem above: A straight line that connects $P_1 = (x_1, y_1)$ and $P_2 = (x_2, y_2)$ crosses the elliptic curve at a third point $-P_3 = -(P_1 \oplus P_2) = (x_3, -y_3)$ which is the negative of the **sum point** $P_3 = P_1 \oplus P_2 = (x_3, y_3)$. In fact, since elliptic curves are symmetric with respect to the x-axis, here $-P_3$ is the point on the curve which is the reflection of P_3 in the x-axis.

If $P_1 = P_2$, then $P_1 \oplus P_2 = 2P_1$ (**point doubling**). The tangent line to the elliptic curve at P_1 crosses the curve again at $-2P_1$, and reflection of this point across the x-axis gives us $2P_1$, which is on the curve as well.

If $P_1 \neq P_2$ and $x_1 = x_2$, then $P_1 \oplus P_2 = \mathcal{O}$ since the line connecting the points P_1 and P_2 is vertical and does not intersect the curve anywhere else, i.e. we can say that it meets the curve again only at infinity. If $P_1 = P_2$ and $y_1 = 0$, then $P_1 \oplus P_2 = 2P_1 = \mathcal{O}$ since the tangent line to the curve at $P_1 = (x_1, 0)$ is vertical.

The geometric descriptions of point addition and point doubling yield some algebraic formulas. Let $\mathcal{E}$ be an elliptic curve defined by $y^2 = x^3 + ax + b$. Let $P_1 = (x_1, y_1)$ and $P_2 = (x_2, y_2)$ be two points on $\mathcal{E}$ with $P_1, P_2 \neq \mathcal{O}$ and $P_1 \neq -P_2$. And let $P_1 + P_2 = P_3 = (x_3, y_3)$. Then,

$$\begin{aligned} x_3 &= m^2 - x_1 - x_2, \\ y_3 &= m \times (x_1 - x_3) - y_1, \end{aligned} \tag{2.4}$$

where

$$m = \begin{cases} \dfrac{y_2 - y_1}{x_2 - x_1}, & \text{if} \quad P_1 \neq P_2 \text{ (and } x_1 \neq x_2), \\ \dfrac{3 \times x_1^2 + a}{2y_1}, & \text{if} \quad P_1 = P_2 \text{ (and } y_1 \neq 0). \end{cases} \tag{2.5}$$

Here m is the slope of the straight line that connects P_1 and P_2. In case $P_1 = P_2$, the slope m is equal to the evaluation of the derivative of the elliptic curve equation $y^2 = x^3 + ax + b$ at point P_1.

Theorem 28 *The point addition on an elliptic curve $\mathcal{E}$ satisfies the following properties:*

(i) (Closure) For any P_1 and P_2 on $\mathcal{E}$, $P_1 \oplus P_2$ also is a point on $\mathcal{E}$.
(ii) (Associativity) $(P_1 \oplus P_2) \oplus P_3 = P_1 \oplus (P_2 \oplus P_3)$ for all P_1, P_2, P_3 on $\mathcal{E}$.
(iii) (Existence of identity) $P \oplus \mathcal{O} = \mathcal{O} + P = P$ for all P on $\mathcal{E}$.
(iv) (Existence of inverses) For any P there exists $-P$ on $\mathcal{E}$ such that $P \oplus (-P) = (-P) \oplus P = \mathcal{O}$.
(v) (Commutativity) $P_1 \oplus P_2 = P_2 \oplus P_1$ for all P_1, P_2 on $\mathcal{E}$.

In other words, $(\mathcal{E}, \oplus)$is an additive abelian group with identity $\mathcal{O}$.

Proof. (i) follows from Bézout's Theorem (Theorem 27) and symmetry of elliptic curves. (ii) has a messy proof; we refer the reader to [27] pp. 20-34. (iii) follows from the definition of $\mathcal{O}$. (iv) follows from the symmetry of elliptic curves. (v) follows from the fact that the straight line through P_1 and P_2 is the same as the line through P_2 and P_1.

It is important to observe that if the elliptic curve $y^2 = x^3 + ax + b$ is over a field K, i.e. $a, b \in K$, and if P_1 and P_2 have coordinates in K, then $P_1 \oplus P_2$ and $2P_1$ also have coordinates in K. Then the following result follows immediately.

Corollary 7 *Let $\mathcal{E} : y^2 = x^3 + ax + b$ be an elliptic curve over a field K. Then the set of points $\mathcal{E}(K)$ on $\mathcal{E}$ with coordinates in K is a subgroup of the group of all points of $\mathcal{E}$.*

2.4.2 Elliptic Curve Point Multiplication

The elliptic curve point multiplication operation takes an integer k and a point P on the curve $\mathcal{E}$, and computes

$$[k]P = P \oplus P \oplus \cdots \oplus P,$$

where we have k summands. The order of an element P is the smallest integer k such that $[k]P = \mathcal{O}$. The order of any point divides the order of the group $(\mathcal{E}, \oplus)$ by Lagrange's Theorem (Theorem 24). If the order of the element P is equal to the group order, then P is called a primitive element.

Given P and Q on $\mathcal{E}$ such that $Q = [k]P$, the computation of the integer k is called the **Elliptic Curve Discrete Logarithm Problem (ECDLP)**. While it is easy to compute Q, it is hard to compute k from $Q = [k]P$. ECDLP is much more difficult than the DLP in $\mathbb{Z}_p^*$.

2.4.3 Elliptic Curves over Finite Fields

Elliptic curves in the short Weierstrass form over finite fields are very useful in cryptography. For cryptographic use, we use the elliptic curves over GF(p), or $\mathbb{Z}_p$,for $p \neq 2, 3$, in the short Weierstrass form

$$\mathcal{E} : y^2 \equiv x^3 + ax + b \quad (2.6)$$

together with a point of infinity $\mathcal{O}$. The discriminant of the curve is $\Delta = -16 \times (4 \times a^3 + 27 \times b^2) \not\equiv 0 \pmod{p}$. Be careful that all the solutions are computed and all the field operations are performed modulo p.

We will denote the group of points of $\mathcal{E}$ by $\mathcal{E}(\mathbb{Z}_p)$, and the order of the group is denoted by $\#\mathcal{E}(\mathbb{Z}_p)$ or $\#\mathcal{E}$. Assigning a particular value of $x \in \mathbb{Z}_p$ in the right-hand side of the equation, namely $z = x^3 + ax + b$, we solve for the quadratic equation

$$y^2 \equiv z \pmod{p}.$$

The solution of this quadratic equation gives the values y and $-y$, since $y^2 = (-y)^2$, implying that the pairs (x,y) and $(x,-y)$ are on the curve. It turns out that if (x,y) is a point on the curve, then $(x,-y)$ must be so as well. This is verified geometrically by the elliptic curves in the form (2.6) being symmetric with respect to the x-axis. Due to this symmetry, since $GF(p)$ has p elements, $\mathscr{E}$ cannot have more than $2 \times p$ points. Including the point at infinity, the order is bounded by

$$\#\mathscr{E} \leq 2 \times p + 1.$$

A more precise bound was found by Hasse.

Theorem 29 ***(Hasse's Theorem)*** *The order of an elliptic curve group $\mathscr{E}$ over $GF(p)$ is bounded by*

$$p + 1 - 2 \times \sqrt{p} \leq \#\mathscr{E} \leq p + 1 + 2 \times \sqrt{p}.$$

On the other hand, although finding the group order of a curve is an elaborate task, it can be computed in polynomial time for any elliptic curve and any finite field. For the setting of ECDLP-based cryptosystems, it is important to know the group order to figure out how secure the group is.

Note that the point multiplication operation which is the computation of $[k]P$ effectively gives

$$[k]P = [k \pmod{p}]P.$$

We further have

$$[k]P \oplus [m]P = [k + m \pmod{p}]P$$
$$[k][m]P = [k \times m \pmod{p}]P.$$

Example 20 *Consider the elliptic curve*

$$\mathscr{E} : y^2 = x^3 + 2x + 3.$$

over the field $\mathbb{Z}_7$. There are six points in the group $\mathscr{E}(\mathbb{Z}_7)$ given by

$$\mathscr{O}, (2,1), (2,6) = (2,-1), (3,1), (3,6) = (3,-1), \text{ and } (6,0),$$

which come in pairs (x,y) and $(x,-y)$ except $(6,0)$. To exemplify the point addition, let's compute the sum $(x_1,y_1) \oplus (x_2,y_2) = (2,1) \oplus (3,1)$. Since $x_1 \neq x_2$, first use the first formula in (2.5) to find m:

$$
\begin{aligned}
m &\equiv (y_2 - y_1) \times (x_2 - x_1)^{-1} \pmod 7 \\
&\equiv (1-1) \times (3-2)^{-1} \pmod 7 \\
&\equiv 0
\end{aligned}
$$

Now we use the value of $m = 0$ to compute x_3 and y_3 by the formulas in (2.4):

$$
\begin{aligned}
x_3 &\equiv m^2 - x_1 - x_2 \pmod 7 \\
&\equiv 0^2 - 2 - 3 \pmod 7 \\
&\equiv 2, \\
y_3 &\equiv m \times (x_1 - x_3) - y_1 \pmod 7 \\
&\equiv 0 \times (2-2) - 1 \pmod 7 \\
&\equiv 6.
\end{aligned}
$$

Therefore, we obtain $(x_3, y_3) = (2,1) \oplus (3,1) = (2,6)$.

To exemplify the point doubling, lets calculate $P_3 = (2,6) \oplus (2,6)$. Since $x_1 = x_2$ and $y_1 = y_2$, we use the second formula in (2.5):

$$
\begin{aligned}
m &\equiv (3 \times {x_1}^2 + a) \times (2 \times y_1)^{-1} \pmod 7 \\
&\equiv (3 \times 2^2 + 2) \times (2 \times 6)^{-1} \pmod 7 \\
&\equiv 0 \times 3 \pmod 7 \\
&\equiv 0
\end{aligned}
$$

Now we use the value of $m = 0$ to compute x_3 and y_3 by the formulas in (2.4), and we obtain $x_3 = 3$ and $y_3 = 1$. Thus, we have $(x_3, y_3) = (2,6) \oplus (2,6) = (3,1)$.

Via point addition and point doubling of the points, we have the following table which shows that the points on the given curve form a group $\mathscr{E}(\mathbb{Z}_7)$.

$\oplus$	$\mathcal{O}$	(2,1)	(2,6)	(3,1)	(3,6)	(6,0)
$\mathcal{O}$	$\mathcal{O}$	(2,1)	(2,6)	(3,1)	(3,6)	(6,0)
(2,1)	(2,1)	(3,6)	$\mathcal{O}$	(2,6)	(6,0)	(3,1)
(2,6)	(2,6)	$\mathcal{O}$	(3,1)	(6,0)	(2,1)	(3,6)
(3,1)	(3,1)	(2,6)	(6,0)	(3,6)	$\mathcal{O}$	(2, 1)
(3,6)	(3,6)	(6,0)	(2,1)	$\mathcal{O}$	(3,1)	(2,6)
(6,0)	(6,0)	(3,1)	(3,6)	(2,1)	(2,6)	$\mathcal{O}$

As the table shows, $\mathcal{O} + P = P + \mathcal{O} = P$ for any P and every point has an inverse $P + (-P) = \mathcal{O}$, such as $(2,1) + (2,6) = \mathcal{O}$.

Chapter 3
Rivest-Shamir-Adleman Algorithm

The RSA algorithm was invented by three MIT professors, Ronald Rivest, Adi Shamir and Leonard Adleman in the summer of 1977 [21], meeting the challenge put forward by the Stanford team Whitfield Diffie, Martin Hellman and Ralph Merkle in 1976. The RSA algorithm was the first public-key cryptographic algorithm, and by far it is also the most deployed. While there are competing algorithms, it remains highly secure and fast, and commonly used. Its security depends on the **eth Root Problem** (or RSA Root Problem), which is described as finding the eth root of a number modulo n where n is factored into two nearly equal-sized prime numbers, as $n = p \times q$.

3.1 Key Generation

Suppose that we use an N-letter alphabet.

(i) Choose positive integers s and t such that $s < t$. The plaintext consists of s-letter blocks, which are regarded as s-digit base-N integers, and the encrypted text consists of t-letter blocks, which are regarded as t-digit base-N integers.
(ii) Choose two sufficiently large primes p and q that differ in length by a few digits in such a way that $N^s < n < N^t$, where $n = p \times q$. In other words $s < \log_N n < t$.
(iii) Choose an integer e such that $\gcd(e, \phi(n)) = 1$, where ϕ is Euler's totient function.
(iv) Compute $d \equiv e^{-1} \pmod{\phi(n)}$.

The pair (n, e) is the public encryption key pair, and the private decryption key d and related parameters are kept secret.

Ç. K. Koç et al., *Partially Homomorphic Encryption*,
https://doi.org/10.1007/978-3-030-87629-6_3

Algorithm 9 Key generation algorithm for the RSA cryptosystem

1: **Input:** Large odd distinct primes p, q.
2: **Output:** Public key (n, e) and secret key d.
3: $n := p \times q$
4: $\phi := (p-1) \times (q-1)$ ▷ Calculates $\phi(n)$.
5: **repeat**
6: e:=RANDINT$(1, \phi - 1)$ ▷ Chooses encryption key.
7: **until** GCD(ϕ, e)=1
8: d:=INVERSE(e, ϕ) ▷ Calculates decryption key.
9: **return** (n, e) and d

3.2 Encryption

After converting the message M (unpadded plaintext) into an integer m (the padded plaintext) using a padding scheme, the sender computes the ciphertext c:

$$c \equiv m^e \pmod{n}.$$

This can be done in a reasonable amount of time using modular exponentiation even if e and n are large integers.

Algorithm 10 Encryption algorithm for the RSA cryptosystem

1: **Input:** Message m, encryption key (n, e).
2: **Output:** Encrypted message c.
3: c:=LRPOW(m, e, n)
4: **return** c

3.3 Decryption

In this part we will use the following fact.

Proposition 3 *For any* $m \in \mathbb{Z}_n$

$$m^{e \times d} \equiv m \pmod{n}, \tag{3.1}$$

where $n = p \times q$ *and* $d^{-1} \equiv e \pmod{\phi(n)}$.

Proof. In fact Euler's Theorem (Theorem 11) already proves this fact for the case when $\gcd(m, n) = 1$ but it works even if $\gcd(m, n) \neq 1$ when n is a product of two distinct primes. First, note that we may write

$$m^{e \times d} \equiv m^{\phi(n) \times j + 1} \pmod{n},$$

where $j \in \mathbb{Z}$, and

$$e \times d - 1 = a \times (p-1) = b \times (q-1)$$

for some integers a and b, since $\phi(p \times q)$ is divisible by both $p-1$ and $q-1$. By the Chinese Remainder Theorem (Theorem 15) it suffices to show that

$$m^{e \times d} \equiv m \pmod{p} \text{ and } m^{e \times d} \equiv m \pmod{q}.$$

Both $m^{e \times d} \equiv m \pmod{p}$ and $m^{e \times d} \equiv m \pmod{q}$ can be considered in two cases. Assuming $m \equiv 0 \pmod{p}$ gives $m^{e \times d} \equiv 0 \pmod{p}$. If $m \not\equiv 0 \pmod{p}$, i.e. $\gcd(m, p) = 1$, then

$$m^{e \times d} \equiv m^{e \times d - 1} m \equiv m^{a \times (p-1)} \times m \equiv (m^{p-1})^a \times m \equiv 1^a \times m \equiv m \pmod{p}$$

as $m^{p-1} \equiv 1 \pmod{p}$ by Fermat's Little Theorem (Theorem 8). Similarly, using $m^{q-1} \equiv 1 \pmod{q}$, one can show that $m^{e \times d} \equiv m \pmod{q}$. This completes the proof of (3.1).

The sender has access to the receiver's public-key pair (n, e) and encrypts her message m to obtain $c = m^e \pmod{n}$ and sends the ciphertext c to the receiver.

The receiver, as the owner of the public-key pair (n, e), has access to the decryption exponent d, among all the secret parameters, $p, q, \phi(n)$. Once the receiver gets the ciphertext c, the padded plaintext m can be recovered using the private decryption key d by performing an exponentiation $m = c^d \pmod{n}$. This holds due to (3.1):

$$c^d \equiv (m^e)^d \equiv m^{e \times d} \equiv m \pmod{n}.$$

Finally the receiver uses the padding scheme to recover the original message M (unpadded plaintext).

Algorithm 11 Decryption algorithm for the RSA cryptosystem

1: **Input:** Received message c, decryption key d.
2: **Output:** Original message m.
3: m:=LRPOW(c, d, n)
4: **return** m

3.4 Homomorphic Properties

Let (n, e) be the public-key pair and $n = p \times q$. Then, the encryption function can be described as:

$$\begin{aligned} E : \mathbb{Z}_n &\to \mathbb{Z}_n \\ m &\mapsto m^e. \end{aligned}$$

Denote the decryption function by D. Consider two padded plaintexts m_1 and m_2. Then

$$\begin{aligned} E(m_1) \times E(m_2) &= m_1^e \times m_2^e \pmod n \\ &\equiv (m_1 \times m_2)^e \pmod n \\ &= E(m_1 \times m_2). \end{aligned}$$

Hence

$$D(E(m_1) \times E(m_2)) = D(E(m_1 \times m_2)). \tag{3.2}$$

On the other hand, denoting addition over $\mathbb{Z}_n$ by $+$, there is no additively homomorphic operation $\boxplus$ known over $\mathbb{Z}_n$ such that

$$D(E(m_1 + m_2)) = D(E(m_1) \boxplus E(m_2))$$

holds. Apparently, RSA is only multiplicatively homomorphic.

However, RSA is not secure unless messages are randomly padded. When m_1 and m_2 are randomly padded as m_1' and m_2', it is impossible to recover $m_1 \times m_2$ from $m_1' \times m_2'$. Therefore, as described, RSA is not often used for its homomorphic property.

3.5 Security

The security of RSA relies on the RSA Root Problem, which is described as finding eth roots modulo $n = p \times q$ (product of two distinct primes) when the factorization of n is unknown.

If an adversary A can factor n and recover p and q, she can also compute $\phi(n)$, since Fact$[n]$ is equivalent to the problem of finding $\phi(n)$ by Theorem 10. Hence A can find the private key d and decrypt the ciphertext. So factoring n is sufficient to break RSA. However, it is unknown whether breaking the RSA cryptosystem is equivalent to the Integer Factorization Problem. In [3] Boneh and Venkatesan showed that breaking low-exponent RSA is not as hard as factoring integers, unless factoring is easy.

The RSA algorithm is malleable due to its homomorphic property. If the adversary A has access to two encrypted messages $E(m_1)$ and $E(m_2)$, then by Eq. (3.2), she can generate another encrypted message as follows:

$$E(m_1) \times E(m_2) = E(m_1 \times m_2).$$

3.6 Example

Suppose that we use a 26-letter alphabet for both plaintext and ciphertext. As a padding scheme, we enumerate letters of the English alphabet modulo 26 starting from the letter A. Let plaintext message units be pairs of letters and ciphertext message units be three letters, i.e. $s = 2$ and $t = 3$. Also let $n = p \times q$ such that

$$26^2 = 676 < n < 26^3 = 17576.$$

Hence $p = 103$, $q = 101$, $n = 10403$ works very well for such an alphabet. This gives $\phi(n) = 10200$. Assume that $e = 1331$ and $d = 1571$, and thus, $1331 \times 1571 \equiv 1 \pmod{5100}$. The public-key pair is $(n, e) = (10403, 1331)$ and $d = 1571$ is the private decryption key. On the other hand, suppose that the message to be sent is the plaintext "REST". First we write the plaintext as two-letter blocks:

$$\begin{aligned} RE &\rightarrow 17 \times 26^1 + 4 \times 26^0 = 446 \\ ST &\rightarrow 18 \times 26^1 + 19 \times 26^0 = 487. \end{aligned}$$

Then we compute $446^{1331} \equiv 9746 \pmod{10403}$ and $487^{1331} \equiv 9790 \pmod{10403}$. Also we write these two numbers as three-digit base-26 numbers and convert them into encrypted three-letter blocks:

$$\begin{aligned} 9746 &= 14 \times 26^2 + 10 \times 26^1 + 22 \times 26^0 \rightarrow OKW \\ 9790 &= 14 \times 26^2 + 12 \times 26^1 + 14 \times 26^0 \rightarrow OMO. \end{aligned}$$

The encrypted text "OKWOMO" is sent to the receiver. Once the message arrives, the receiver groups it into three-letter blocks and computes

$$\begin{aligned} OKW &\rightarrow 14 \times 26^2 + 10 \times 26^1 + 22 \times 26^0 = 9746 \\ OMO &\rightarrow 14 \times 26^2 + 12 \times 26^1 + 14 \times 26^0 = 9790. \end{aligned}$$

Then the message m is recovered by using the decryption key $d = 1571$ as follows:

$$\begin{aligned} 9746^{1571} &\equiv 446 \pmod{n} = 17 \times 26^1 + 4 \times 26^0 \rightarrow RE \\ 9790^{1571} &\equiv 487 \pmod{n} = 18 \times 26^1 + 19 \times 26^0 \rightarrow ST. \end{aligned}$$

Finally the two-letter blocks are combined and the original text "REST" is recovered. As an example to illustrate how multiplicative homomorphism E in Eq. (3.2) works, consider two numerical messages $m_1 = 446$ and $m_2 = 487$. Then

$$\begin{aligned} E(m_1 \times m_2) &= E(446 \times 487) = E(9142) \\ &\equiv 9142^{1331} \pmod{10403} \\ &\equiv 7427 \pmod{10403} \\ E(m_1 \times m_2) &\equiv E(m_1) \times E(m_2), \end{aligned}$$

Chapter 4
Goldwasser-Micali Algorithm

The Goldwasser–Micali (GM) algorithm [9] is an asymmetric-key encryption algorithm developed by Shafi Goldwasser and Silvio Micali in 1982. The GM algorithm has the distinction of being the first probabilistic public-key encryption scheme, where each plaintext has several corresponding ciphertexts. This stems from an additional random parameter chosen at the encryption step. However, it is not an efficient algorithm. The expansion rate defined as the ratio between the lengths of the ciphertext and the plaintext is large. For this reason, it is not widely deployed. Still, it can inspire construction of efficient cryptosystems that are useful in the real world. The security of the GM algorithm is based on the Quadratic Residuosity Problem ($\mathrm{QR}[n]$) modulo $n = p \times q$ (see Sect. 2.1.10).

4.1 Key Generation

For key generation:

(i) Choose two random large primes p and q, and compute $n = p \times q$.

(ii) Choose a quadratic non-residue $x \in \mathbb{Z}_n^*$ with Jacobi symbol $\left(\dfrac{x}{n}\right) = 1$, i.e. a pseudo-square. This choice is accomplished by finding $x \in \mathbb{Z}_n^*$ such that $\left(\dfrac{x}{p}\right) = \left(\dfrac{x}{q}\right) = -1$. That is, we find a quadratic non-residue modulo both p and q, since

$$\left(\frac{x}{n}\right) = \left(\frac{x}{p}\right) \times \left(\frac{x}{q}\right) = (-1) \times (-1) = 1.$$

Ç. K. Koç et al., *Partially Homomorphic Encryption*,
https://doi.org/10.1007/978-3-030-87629-6_4

By choosing p and q as Blum integers, i.e. $p \equiv 3 \pmod 4$ and $q \equiv 3 \pmod 4$, the integer $n-1$ is guaranteed to be a quadratic non-residue with $\left(\frac{n-1}{n}\right) = \left(\frac{-1}{n}\right) = 1$.

The public key consists of (n,x) and the private key is (p,q).

Algorithm 12 Key generation algorithm for the Goldwasser-Micali cryptosystem

1: **Input:** Large random primes p,q.
2: **Output:** Public key (n,x) and private key (p,q).
3: $n := p \times q$
4: **repeat**
5: $\quad x$:=RANDINT$(1,n-1)$
6: **until** GCD(x,n)=1 and JAC(x,p)=-1 and JAC(x,q)=-1
$\quad\quad$ ▷ Choose a quadratic non-residue x in $\mathbb{Z}_n^*$ with Jacobi symbol $(x/n) = 1$.
7: **return** (n,x) and (p,q)

4.2 Encryption

The plaintext is a string of bits $(m_1, m_2, \ldots, m_k)$. The sender picks uniformly at random $y_i \in \mathbb{Z}_n^*$ for each bit m_i, and encrypts one bit of information at a time by computing

$$c_i \equiv y_i^2 \times x^{m_i} \pmod n.$$

The ciphertext generated is $(c_1, c_2, \ldots, c_k)$, such that $c_i \in \mathbb{Z}_n$ for $i = 1, 2, \ldots, k$. In other words, every single bit of the plaintext is encrypted to a $\log_2 n$-bit integer, and thus, the k-bit plaintext is encrypted to $k \times \log_2 n$ bits. Since its security depends on the factorization of integer n, we have $\log_2 n > 1024$. The GM algorithm is highly inefficient.

Algorithm 13 Encryption algorithm for the Goldwasser-Micali cryptosystem

1: **Input:** Message string of bits $(m_1, m_2, \ldots, m_k)$, encryption key (n,x).
2: **Output:** Ciphertext string $(c_1, c_2, \ldots, c_k)$.
3: **for** $i = 1$ **to** k **do**
4: $\quad$ **repeat**
5: $\quad\quad y_i$:=RANDINT$(1,n)$
6: $\quad$ **until** GCD(y_i,n)=1
7: $\quad c_i := y_i^2 \times x^{m_i} \pmod n$
8: **return** $(c_1, c_2, \ldots, c_k)$

4.3 Decryption

To decrypt the message and get the plaintext back, the receiver determines whether c_i is a quadratic residue modulo n for $i = 1, \ldots, k$. If c_i is a quadratic residue modulo n, then $m_i = 0$; if c_i is a quadratic non-residue modulo n, then $m_i = 1$.

Since p and q are known by the receiver (the owner of the public key), she can decide the quadratic residuosity of c_i modulo p and modulo q by computing the Legendre symbols. If c_i is a quadratic residue modulo both p and q, then c_i is a quadratic residue modulo n by Theorem 18.

Algorithm 14 Decryption algorithm for the Goldwasser-Micali cryptosystem

1: **Input:** Received message string $(c_1, c_2, \ldots, c_k)$, decryption key (p, q).
2: **Output:** Original message string $(m_1, m_2, \ldots, m_k)$.
3: **for** $i = 1$ **to** k **do**
4: **if** JAC(c_i, p)=1 and JAC(c_i, q)=1 **then** $m_i = 0$
5: **else** $m_i = 1$
6: **return** $(m_1, m_2, \ldots, m_k)$

4.4 Homomorphic Properties

The bitwise encryption function of the Goldwasser-Micali algorithm is

$$\begin{aligned} E : (\{0,1\}, \oplus) &\to (\mathbb{Z}_n^*, \times) \\ m &\mapsto y^2 \times x^m, \end{aligned}$$

where $\oplus$ denotes addition modulo 2. If c and c' are the encryptions of bits m and m', then

$$\begin{aligned} E(m) &= c \equiv y^2 \times x^m \pmod{n} \\ E(m') &= c' \equiv (y')^2 \times x^{m'} \pmod{n}, \end{aligned}$$

where y and y' are randomly selected integers in $\mathbb{Z}_n^*$. Denote the decryption function by D. Then the homomorphic property is

$$\begin{aligned} E(m) \times E(m') &\equiv (y^2 \times x^m) \times ((y')^2 \times x^{m'}) \pmod{n} \\ &\equiv (y \times y')^2 \times x^{m \oplus m'} \pmod{n} \\ &\equiv E(m \oplus m'), \end{aligned}$$

where the randomness in the encryption of $m \oplus m'$ is $y \times y'$, which is neither uniformly distributed in $\mathbb{Z}_n^*$ nor independent of the randomness in $E(m)$ and $E(m')$.

However this can be addressed by re-randomization, which is explained at the end of this section.

It turns out that

$$D(E(m) \times E(m')) = D(E(m \oplus m')).$$

Therefore, the GM algorithm is additively homomorphic. However, there is no multiplicative homomorphic property of E. In other words, there does not exist an operation $\boxtimes$ in the ciphertext space such that

$$D(E(m \times m')) = D(E(m) \boxtimes E(m')).$$

The GM algorithm, however, is multiplicatively and additively homomorphic with an integer scalar k.

For an integer k, we have

$$E(m)^k \equiv (y^2 \times x^m)^k \pmod{n} \equiv (y^k)^2 \times x^{k \times m} \pmod{n} \equiv E(k \times m)$$

with the randomness $y^k \in \mathbb{Z}_n^*$. Hence

$$D(E(m)^k) = D(E(k \times m)).$$

Similarly, for an integer scalar k, we have

$$E(m+k) \equiv y^2 \times x^{m+k} \pmod{n} \equiv y^2 \times x^m \times x^k \pmod{n} \equiv E(m) \times (x^k \pmod{n}),$$

which means

$$D(E(m+k)) = D(E(m) \times x^k).$$

Re-randomization is possible for the GM algorithm. Let $y' \in \mathbb{Z}_n^*$ be a random number and $c = E(m)$. Then,

$$(y')^2 \times E(m) \equiv (y')^2 \times y^2 \times x^m \pmod{n} \equiv (y' \times y)^2 \times x^m \pmod{n},$$

which is a valid encryption of m with the randomness $y' \times y \in \mathbb{Z}_n^*$. Hence

$$D((y')^2 \times c) = D(c).$$

4.5 Security

The security of the Goldwasser-Micali scheme depends on the Integer Factorization Problem, since factoring $n = p \times q$ reveals the private key (p, q).

The concept of semantic security was first introduced by Goldwasser and Micali in their seminal paper [9]. Semantic security differs from perfect security in the sense that the adversary is allowed to have polynomially bounded computing power. A cryptosystem is semantically secure if whatever an adversary can compute about the plaintext given the ciphertext, she can also compute without the ciphertext. Goldwasser and Micali [9] proved that each probabilistic encryption scheme has polynomial security (sometimes called indistinguishability, or IND security), and that polynomial security implies semantic security. Shortly thereafter, Micali, Rackoff and Sloan [15] showed that these two notions coincide.

The semantic security of the GM cryptosystem is based on the QR$[n]$ problem.

Theorem 30 *If QR$[n]$ is intractable, then the Goldwasser-Micali cryptosystem is semantically secure.*

Proof. Assume that the GM cryptosystem is not semantically secure. Then given a ciphertext c, an adversary can decide whether it is an encryption of 0 or 1. If it is an encryption of 0, then c is a quadratic residue modulo n; otherwise, c is a quadratic non-residue modulo n. Therefore, QR$[n]$ is not intractable.

4.6 Example

Consider $p = 5$ and $q = 7$. Compute $n = p \times q = 35$. Choose $x = 3$, which is a quadratic non-residue modulo 35 and also $\left(\frac{3}{35}\right) = \left(\frac{3}{5}\right) \times \left(\frac{3}{7}\right) = (-1) \times (-1) = 1$. Then $(n,x) = (35,3)$ becomes the public encryption key and $(p,q) = (5,7)$ is the private decryption key.

Let $m = (m_1, m_2, m_3) = (1,0,0)$ and choose $y_1 = 4$, $y_2 = 11$, $y_3 = 16$, which are elements of $\mathbb{Z}_{35}^*$. Then we calculate the ciphertext

$$c = E(m) = (E(m_1), E(m_2), E(m_3)) = (c_1, c_2, c_3)$$

as

$$\begin{aligned} c_1 &\equiv y_1^2 \times x^{m_1} \equiv 4^2 \times 3^1 \equiv 13 \pmod{35} \\ c_2 &\equiv y_2^2 \times x^{m_2} \equiv 11^2 \times 3^0 \equiv 16 \pmod{35} \\ c_3 &\equiv y_3^2 \times x^{m_3} \equiv 16^2 \times 3^0 \equiv 11 \pmod{35}, \end{aligned}$$

i.e. $c = (13, 16, 11)$ is the encrypted message. Since we have the factorization of n as the private key, we can test each component (encryption of each bit) of the encrypted message for quadratic residuosity. If a component is a quadratic residue modulo both 5 and 7, then it is decrypted as 0. Otherwise it is decrypted as 1. We start with $c_1 = 13$:

$$\begin{aligned} 13 &\equiv 3 \pmod{5}, \text{which is a quadratic non-residue.} \\ 13 &\equiv 6 \pmod{7}, \text{which is a quadratic non-residue.} \end{aligned}$$

So the decryption of c_1 becomes 1, which is equal to m_1.

$$16 \equiv 1 \pmod 5, \text{which is a quadratic residue.}$$
$$16 \equiv 2 \pmod 7, \text{which is a quadratic residue.}$$

So the decryption of $c_2 = 16$ becomes 0, which is equal to m_2.

$$11 \equiv 1 \pmod 5, \text{which is a quadratic residue.}$$
$$11 \equiv 4 \pmod 7, \text{which is a quadratic residue.}$$

So the decryption of $c_3 = 11$ becomes 0, which is equal to m_3. Hence $m = (1,0,0)$ is recovered.

To illustrate the homomorphic property, further let $m' = (m'_1, m'_2, m'_3) = (1,1,0)$ with $y'_1 = 6, y'_2 = 13, y'_3 = 19$. We compute $c' = E(m') = (c'_1, c'_2, c'_3)$ bit by bit:

$$c'_1 \equiv (y'_1)^2 \times x^{m'_1} \equiv 6^2 \times 3^1 \equiv 3 \pmod{35}$$
$$c'_2 \equiv (y'_2)^2 \times x^{m'_2} \equiv 13^2 \times 3^1 \equiv 17 \pmod{35}$$
$$c'_3 \equiv (y'_3)^2 \times x^{m'_3} \equiv 19^2 \times 3^0 \equiv 11 \pmod{35},$$

i.e. $c' = (3, 17, 11)$. Then

$$c \times c' = E(m) \times E(m') = (13 \times 3, 16 \times 17, 11 \times 11) = (4, 27, 16).$$

Now we decrypt $c \times c'$ bit by bit:

$$4 \equiv 4 \pmod 5, \text{which is a quadratic residue.}$$
$$4 \equiv 1 \pmod 7, \text{which is a quadratic residue.}$$

So 4 is decrypted as 0.

$$27 \equiv 2 \pmod 5, \text{which is a quadratic non-residue.}$$
$$27 \equiv 6 \pmod 7, \text{which is a quadratic non-residue.}$$

So 27 is decrypted as 1.

$$16 \equiv 1 \pmod 5, \text{which is a quadratic residue.}$$
$$16 \equiv 2 \pmod 7, \text{which is a quadratic residue.}$$

So 16 decrypted as 0. Hence $c \times c'$ is decrypted as $(0,1,0) = m \oplus m'$, which verifies the equality $D(E(m \oplus m')) = D(E(m) \times E(m'))$.

To illustrate the multiplicative homomorphic property with an integer scalar, further let $k = 3$ be a constant. We consider the previously used message $(m_1, m_2, m_3) = (1,0,0)$ and its accompanying randomness $y = (y_1, y_2, y_3) = (4, 11, 16)$. Then we compute $E(k \times m) = c^{(k)} = (c_1^{(k)}, c_2^{(k)}, c_3^{(k)})$ bit by bit as

$$c_1^{(k)} \equiv (y_1^k)^2 x^{k\times m_1} \equiv 29^2 3^3 \equiv 27 \pmod{35}$$
$$c_2^{(k)} \equiv (y_2^k)^2 x^{k\times m_2} \equiv 1^2 3^0 \equiv 1 \pmod{35}$$
$$c_3^{(k)} \equiv (y_3^k)^2 x^{k\times m_3} \equiv 1^2 3^0 \equiv 1 \pmod{35},$$

which means $c^{(k)} = (27,1,1)$. Observe that $c^k = (13^3, 16^3, 11^3) = (27,1,1)$. Hence the equality $(E(m))^k = E(k \times m)$ is verified.

To illustrate the additive homomorphic property with an integer scalar, consider previously used k, m and y. Then we compute $E(m+k)$ bit by bit as

$$(y_1)^2 x^{m_1+k} \equiv 4^2 3^4 \equiv 1 \pmod{35}$$
$$(y_2)^2 x^{m_2+k} \equiv 11^2 3^3 \equiv 12 \pmod{35}$$
$$(y_3)^2 x^{m_3+k} \equiv 16^2 3^3 \equiv 17 \pmod{35},$$

which yields $E(m+k) = (1,12,17)$. Observe that $E(m) \times (x^k \pmod{n}) = (13 \times 3^3, 16 \times 3^3, 11 \times 3^3) \pmod{35} = (1,12,17)$, Hence the equality $E(m+k) = E(m) \times (x^k \pmod{n})$ is verified.

To exemplify the re-randomization property, choose $y_1' = 6$, $y_2' = 13$, $y_3' = 19$ and calculate $(y_1')^2 = 1$, $(y_2')^2 = 29$, $(y_3')^2 = 11$. Consider $c''' = (c_1 \times (y_1')^2, c_2 \times (y_2')^2, c_3 \times (y_3')^2) = (13,9,16)$. Now decrypt c''' bit by bit:

$$13 \equiv 3 \pmod{5}, \text{which is a quadratic non-residue.}$$
$$13 \equiv 6 \pmod{7}, \text{which is a quadratic residue.}$$

So the decryption of 13 becomes 1.

$$9 \equiv 4 \pmod{5}, \text{which is a quadratic residue.}$$
$$9 \equiv 2 \pmod{7}, \text{which is a quadratic residue.}$$

So the decryption of 9 becomes 0.

$$16 \equiv 1 \pmod{5}, \text{which is a quadratic residue.}$$
$$16 \equiv 2 \pmod{7}, \text{which is a quadratic residue.}$$

So the decryption of 16 becomes 0. Hence c''' decrypts to $m''' = (1,0,0) = m$, which means c''' is a re-randomized version of ciphertext c.

Chapter 5
ElGamal Algorithm

The ElGamal cryptosystem [6] is a public-key encryption scheme proposed by Taher ElGamal in 1985. He was a Ph.D. student of Martin Hellman, the co-inventor of the Diffie-Hellman key exchange algorithm. The ElGamal cryptosystem essentially turns the Diffie-Hellman key exchange method into an encryption algorithm. The security of the ElGamal algorithm is based on the difficulty of solving the Diffie-Hellman Problem (DHP) in $\mathbb{Z}_p^*$ (see Sect. 2.2.7).

5.1 Multiplicatively Homomorphic ElGamal Algorithm

5.1.1 Key Generation

For key generation:

(i) Choose two random large primes p ($\geq$ 2048 bits) and q ($\geq$ 256 bits) such that $q|(p-1)$.
(ii) Choose a cyclic subgroup G_q of $\mathbb{Z}_p^*$ of order q with generator g, i.e. select some $y \in \mathbb{Z}_p^*$ and compute $g \equiv y^{(p-1)/q} \pmod{p}$.
(iii) Select a random $x \in \mathbb{Z}_q$ and set $h = g^x \pmod{p}$.

The public key is (p, q, g, h) and the private key is x.

5.1.2 Encryption

The plaintext is $m \in G_q$. The sender generates a random number $r \in \mathbb{Z}_q$ and computes the encryption of the plaintext m which is the ciphertext pair $E(m) = (c_1, c_2)$, where

Ç. K. Koç et al., *Partially Homomorphic Encryption*,
https://doi.org/10.1007/978-3-030-87629-6_5

Algorithm 15 Key generation algorithm for the multiplicative ElGamal cryptosystem

1: **Input:** Large odd distinct primes p, q, where $q \mid (p-1)$.
2: **Output:** Public key (p,q,g,h) and private key x.
3: y:=RANDINT$(1, p-1)$
4: g:=LRPOW$(y, (p-1)/q, p)$ ▷ Calculates a generator of G_q .
5: x:=RANDINT$(0, q-1)$ ▷ Selects decryption key.
6: h:=LRPOW(g, x, p)
7: **return** $(p,q,g,h), x$

$$c_1 \equiv g^r \pmod{p}$$
$$c_2 \equiv m \times h^r \pmod{p}.$$

The length of the ciphertext pair is twice the length of the plaintext. This message expansion can be regarded as a slight disadvantage. However, the randomization in encryption is an advantage of the ElGamal scheme.

For encryption of each message, a new r is chosen to be a uniformly random integer in order to ensure security. If r is used more than once, the knowledge of a message m enables the adversary to compute other messages as follows.

Let

$$c_1 \equiv g^r \pmod{p}, \quad c_2 \equiv m_1 \times h^r \pmod{p}$$

and

$$c_1{}' \equiv g^{r'} \pmod{p}, \quad c_2{}' \equiv m_2 \times h^{r'} \pmod{p}.$$

If the same random number is selected in both encryptions, i.e. $r = r'$, then $m_2 = m_1 \times c_2^{-1} \times c_2' = m_1 \times (m_1 \times h^r)^{-1} \times (m_2 \times h^{r'}) \pmod{p}$. Therefore, if m_1 is known, then m_2 is computed easily, provided that $r = r'$.

Algorithm 16 Encryption algorithm for the multiplicative ElGamal cryptosystem

1: **Input:** Message $m \in G_q$, encryption key (p,q,g,h).
2: **Output:** Ciphertext pair (c_1, c_2).
3: r:=RANDINT$(0, q-1)$
4: c_1:=LRPOW(g, r, p)
5: c_2:=$m\times$(LRPOW(h, r, p)) $\pmod{p}$
6: **return** (c_1, c_2)

5.1.3 Decryption

The legitimate receiver holds the private key x and knows the values p, q and g. The receiver can decrypt the ciphertext (c_1, c_2), without knowing the value of r, by computing u_1 and u_2 as

$$u_1 = (g^r)^x = (g^x)^r \equiv h^r \pmod{p}$$
$$u_2 = {u_1}^{-1} \times c_2 \equiv h^{-r} \times (m \times h^r) \equiv m \pmod{p},$$

where ${u_1}^{-1}$ is the multiplicative inverse of u_1 in the group G_q. This inverse can be computed using the Extended Euclidean Algorithm.

Algorithm 17 Decryption algorithm for the multiplicative ElGamal cryptosystem

1: **Input:** Received message pair (c_1, c_2), decryption key x.
2: **Output:** Original message m.
3: u_1:=LRPOW(c_1, x, p)
4: m:=(INVERSE(u_1, p))$\times c_2 \pmod{p}$
5: **return** m

5.1.4 Homomorphic Properties

The encryption function of the multiplicative ElGamal cryptosystem is described by

$$E : (G_q, \times) \to (G_q \times G_q, \times)$$
$$m \mapsto (g^r, m \times h^r),$$

where $\times$ denotes the usual multiplication operation, and $G_q \times G_q$ is the direct product of group G_q with itself. E is homomorphic with respect to multiplication, i.e. $D(E(m_1) \times E(m_2)) = D(E(m_1 \times m_2))$, where D denotes the decryption function. One can check this easily as follows.

The encryptions of m_1 and m_2 are the pairs $E(m_1) = (c_1, c_2)$ and $E(m_2) = ({c_1}', {c_2}')$, respectively, where

$$c_1 \equiv g^r \pmod{p}$$
$${c_1}' \equiv g^{r'} \pmod{p}$$
$$c_2 \equiv m_1 \times h^r \pmod{p}$$
$${c_2}' \equiv m_2 \times h^{r'} \pmod{p}.$$

Here the random numbers r and r' are different. The pairwise products of the ciphertext pairs are

$$c_1 \times c_1' \equiv g^r \times g^{r'} \pmod{p}$$
$$\equiv g^{r+r'} \pmod{p}$$
$$c_2 \times c_2' \equiv (m_1 \times h^r) \times (m_2 \times h^{r'}) \pmod{p}$$
$$\equiv m_1 \times m_2 \times h^{r+r'} \pmod{p}.$$

Therefore, the product of the encrypted messages $E(m_1) \times E(m_2) = (c_1 \times c'_1, c_2 \times c'_2)$ is equal to the encryption of the product of messages $E(m_1 \times m_2)$ as follows:

$$E(m_1) \times E(m_2) \equiv (g^{r+r'}, m_1 \times m_2 \times h^{r+r'}) \pmod{p} \equiv E(m_1 \times m_2),$$

where the randomness in the encryption of $m_1 \times m_2$ is $r + r'$, which is neither uniformly distributed in $\mathbb{Z}_p$ nor independent of the randomness in $E(m_1)$ and $E(m_2)$. However this can be addressed by re-randomization, which is explained at the end of this section.

It turns out that

$$D(E(m_1) \times E(m_2)) = D(E(m_1 \times m_2)).$$

E also preserves exponentiation by an integer scalar. For a random integer scalar k, we have

$$\begin{aligned} E(m)^k &= ({c_1}^k, {c_2}^k) \\ &\equiv ((g^r)^k, (m \times h^r)^k) \pmod{p} \\ &\equiv (g^{r \times k}, m^k \times h^{r \times k}) \pmod{p} \\ &\equiv E(m^k), \end{aligned}$$

which means

$$D(E(m)^k) = D(E(m^k)).$$

One can naturally ask whether there is an additive homomorphic property of the encryption function E. If there is, then one should be able to obtain the encrypted message $E(m_1 + m_2)$ from the ciphertext pairs $E(m_1) = (c_1, c_2)$ and $E(m_2) = (c'_1, c'_2)$. In other words, there must exist an additive operation $\boxplus$ in the ciphertext space such that

$$\begin{aligned} E(m_1 + m_2) &\equiv (g^{r''}, (m_1 + m_2) \times h^{r''}) \pmod{p} \\ &\equiv (g^r, m_1 \times h^r) \boxplus (g^{r'}, m_2 \times h^{r'}) \pmod{p} \\ &\equiv E(m_1) \boxplus E(m_2). \end{aligned}$$

However, we do not know such an operation satisfying this homomorphism. But with a small change in the algorithm, the addition operation can be supported, which yields the **additively homomorphic ElGamal cryptosystem**. This version is the subject of the next section.

Re-randomization is possible for the multiplicative ElGamal algorithm. Let $E(m) = c = (c_1, c_2) \equiv (g^r, m \times h^r) \pmod{p}$ for random $r \in \mathbb{Z}_q$, and $r' \in \mathbb{Z}_q$ be another chosen random number. Then

$$\begin{aligned} (c_1 \times g^{r'}, c_2 \times h^{r'}) &\equiv (g^r \times g^{r'}, m \times h^r \times h^{r'}) \pmod{p} \\ &\equiv (g^{r+r'}, m \times h^{r+r'}) \pmod{p}, \end{aligned}$$

which is a re-randomized ciphertext of the original message m where $r+r' \in \mathbb{Z}_q$. Hence

$$D(c) = D(c \times (g^{r'}, h^{r'})).$$

5.1.5 Security

The security of the multiplicatively homomorphic ElGamal cryptosystem is based on the hardness of the Computational Diffie-Hellman Problem (CDHP) and the Decisional Diffie-Hellman Problem (DDHP) in the underlying group G_q.

Theorem 31 *If CDHP is intractable in G_q, then the multiplicatively homomorphic ElGamal encryption function is not invertible by an adversary.*

Proof. If the CDHP is intractable in G_q, then given g^r and g^x, an adversary cannot calculate $g^{r \times x}$ without any knowledge of r and x. Assume the adversary can decrypt the ciphertext $(c_1, c_2) = (g^r, m \times h^r)$ and obtain the message m. Since the public key $h = g^x$ and $c_1 = g^r$ are known by the adversary, she can calculate $m^{-1}c_2 = h^r = (g^x)^r = g^{x \times r}$, which contradicts the fact that CDHP is intractable in G_q.

Theorem 32 *The multiplicatively homomorphic ElGamal cryptosystem is semantically secure if and only if DDHP is intractable in G_q.*

Proof. ($\Rightarrow$) Assume that DDHP is not intractable in the underlying group G_q of the cryptosystem. Then an adversary can decide the correctness of any D-H triplet by means of a D-H oracle. Let m_0 and m_1 be two different messages known by the adversary. Let $c = (c_1, c_2) = (g^r, m_i \times h^r)$ be the given encryption of either m_0 or m_1, where r is randomly chosen from $\mathbb{Z}_q$. The adversary selects a random $y \in \mathbb{Z}_q$ and computes the triple

$$\begin{aligned}(c_1, h \times g^y, c_2 \times c_1{}^y \times m_0{}^{-1}) &= (g^r, g^x \times g^y, m_i \times h^r \times (g^r)^y \times m_0{}^{-1}) \\ &= (g^r, g^{x+y}, m_i \times g^{x \times r} \times g^{r \times y} \times m_0{}^{-1}) \\ &= (g^r, g^{x+y}, g^{r \times (x+y)} \times m_i \times m_0{}^{-1}).\end{aligned}$$

The adversary can decide whether this is a correct D-H triplet. If $m_i \equiv m_0 \pmod{p}$, then $i = 0$ except with negligible probability. If $m_i \not\equiv m_0 \pmod{p}$, then $i = 1$ except with negligible probability. It turns out that the adversary can distinguish the encryptions of two given messages, which implies the lack of semantic security.

($\Leftarrow$) Conversely, assume that the multiplicatively homomorphic ElGamal cryptosystem is not semantically secure. Then the adversary can distinguish the encryptions of any two given messages, say m_0 and m_1. Let r be a randomly chosen number and k be an integer from $\mathbb{Z}_q$. If $g^k \equiv g^{x \times r} \pmod{p}$, then $c = (g^r, m_0 \times g^k)$ would be a valid encryption of m_0. Since the adversary has the ability to distinguish the encryptions of given messages, she can decide whether the triplet (g^r, g^x, g^k) is a D-H triplet or not. Hence the DDHP is solved by the adversary.

On the other hand, ElGamal encryption is unconditionally malleable and so is not secure under chosen ciphertext attack. Given an encryption $(c_1, c_2) = (g^r, m \times h^r)$ of some message m, one can create $(c_1, t \times c_2) = (g^r, t \times m \times h^r)$, which is a valid encryption of $t \times m$ for any t, without knowing m, nor the random r, nor the secret key x.

5.1.6 Example

Consider primes $p = 307$ and $q = 17$. Note that 17 divides $306 = 307 - 1$. The related multiplicative group is

$$\mathbb{Z}_p^* = \mathbb{Z}_{307}^* = \{1, 2, \ldots, 306\}.$$

Choose $g = 272$ with $\text{ord}(272) = 17 = q$. The plaintext space is

$$\begin{aligned} G_q &= \langle 272 \rangle \\ &= \{272, 304, 105, 9, 299, 280, 24, 81, 235, 64, 216, 115, 273, 269, 102, 114, 1\}. \end{aligned}$$

Next we choose the private key as $x = 4 \in \mathbb{Z}_{17}$ and one of the public key parameters h becomes

$$h \equiv g^x \equiv 272^4 \equiv 9 \pmod{307}.$$

Let $m = 24 \in G_q$ and choose $r = 5$. Then

$$\begin{aligned} E(m) &= (c_1, c_2) \\ &= (g^r, m \times h^r) \pmod{p} \\ &\equiv (272^5, 24 \times 9^5) \pmod{307} \\ &= (299, 64) \end{aligned}$$

is the encrypted message. Once we have $(c_1, c_2) = (299, 64)$, we calculate

$$u_1 = 299^4 \equiv 105 \pmod{307}.$$

Then $u_1^{-1} \equiv 269 \pmod{307}$ and the message

$$m = u_2 = u_1^{-1} c_2 \equiv 269 \times 64 \equiv 24 \pmod{307}$$

is recovered.

To illustrate the homomorphic property, further let $m' = 64 \in G_q$ with $r' = 8$. Then

$$\begin{aligned} E(m') &= (c_1', c_2') \\ &= (g^{r'}, m' \times h^{r'}) \pmod p \\ &\equiv (272^8, 64 \times 9^8) \pmod{307} \\ &= (81, 81). \end{aligned}$$

Compute

$$c_1 \times c_1' \equiv 299 \times 81 \equiv 273 \pmod{307}$$

and

$$c_2 \times c_2' \equiv 64 \times 81 \equiv 272 \pmod{307}.$$

So $E(m) \times E(m') = (c_1 \times c_1', c_2 \times c_2') = (273, 272)$. On the other hand, consider the message $m \times m'$ with randomness $r + r'$.

$$\begin{aligned} E(m \times m') &= E(24 \times 64 \pmod{307}) \\ &= E(1) \\ &= (g^{r+r'}, h^{r+r'}) \pmod p \\ &\equiv (272^{5+8}, 9^{5+8}) \pmod{307} \\ &= (273, 272). \end{aligned}$$

Therefore, $E(m \times m') = E(m) \times E(m')$.

Further let $k = 3$. Then

$$E(m) = (c_1, c_2) = (299, 64) \pmod{307}$$

and

$$E(m)^k \equiv (c_1^k, c_2^k) \equiv (299^3, 64^3) \equiv (102, 273) \pmod{307}.$$

On the other hand, $m^k \equiv 24^3 \equiv 9 \pmod{307}$. Encryption of m^k with randomness $r \times k$ is

$$\begin{aligned} E(m^k) \equiv (g^{r \times k}, m^k \times h^{r \times k}) \pmod p &\equiv (272^{5\times3}, 9 \times 9^{5\times3}) \pmod{307} \\ &\equiv (102, 273) \pmod{307}. \end{aligned}$$

Therefore, the equality $E(m)^k = E(m^k)$ is verified.

To exemplify the re-randomization property, choose $r'' = 4 \in \mathbb{Z}_q$. We already have $E(m) = (c_1, c_2) = (299, 64)$. Re-randomizing the ciphertext, we get

$$(c_1 \times g^{r''}, c_2 \times h^{r''}) \equiv (299 \times 272^4, 64 \times 9^4)$$
$$\equiv (299 \times 9, 64 \times 114)$$
$$\equiv (235, 235) \pmod{307}.$$

Decryption of this ciphertext yields the following equivalences.

$$u_1 \equiv (235)^x \equiv 235^4 \equiv 304 \pmod{307},$$
$$u_1^{-1} \equiv 304^{-1} \equiv 102 \pmod{307},$$
$$u_2 \equiv u_1^{-1} c_2 \equiv 102 \times 235 \equiv 24 \pmod{307},$$

which proves that $(235, 235)$ is a valid encryption of $m = 24$.

5.2 Additively Homomorphic ElGamal

All the domain parameters and key-pair generation are the same as in the multiplicative version of ElGamal. But here we first transform the plaintext $m \in \mathbb{Z}_q$ into the group element g^m before encryption.

5.2.1 Encryption

The plaintext is $m \in \mathbb{Z}_q$. The sender generates a random number $r \in \mathbb{Z}_q$ and computes the encryption of the plaintext m, which is the ciphertext pair $E(m) = (c_1, c_2)$, where

$$c_1 \equiv g^r \pmod{p}$$
$$c_2 \equiv g^m \times h^r \pmod{p}.$$

For encryption of each message, a new r is chosen to ensure security as explained in the encryption part of the multiplicative version of ElGamal.

Algorithm 18 Encryption algorithm for the additive ElGamal cryptosystem

1: **Input:** Message $m \in \mathbb{Z}_q$, encryption key (p, q, g, h).
2: **Output:** Ciphertext pair (c_1, c_2).
3: r:=RANDINT$(1, q)$
4: c_1:=LRPOW(g, r, p)
5: c_2:=(LRPOW(g, m, p))×(LRPOW(h, r, p)) $\pmod{p}$
6: **return** (c_1, c_2)

5.2.2 Decryption

The legitimate receiver holds the private key x and knows the values p, q and g. So the receiver can decrypt the ciphertext (c_1, c_2), without knowing the value of r, by computing u_1 and u_2 as

$$u_1 = (g^r)^x = (g^x)^r \equiv h^r \pmod{p}$$
$$u_2 = {u_1}^{-1} \times c_2 \equiv h^{-r} \times (g^m \times h^r) \equiv g^m \pmod{p},$$

where ${u_1}^{-1}$ is the modular multiplicative inverse of u_1 in the group G_q. So the owner of the private key obtains g^m after decryption, but cannot properly decrypt the ciphertext pair (c_1, c_2) and cannot obtain m. If m is from a small set, i.e. q is small, then m can be recovered by an exhaustive search. However, in the key generation step q is chosen to be a large prime. So, in order to find m from g^m, the receiver needs to solve a Discrete Logarithm Problem, which is intractable for the group from which g is chosen. Therefore this version of ElGamal is not practical.

5.2.3 Homomorphic Properties

The encryption function of the additive ElGamal cryptosystem is described by

$$E : (\mathbb{Z}_q, +) \rightarrow (G_q \times G_q, \times)$$
$$m \mapsto (g^r, g^m \times h^r),$$

where $\times$ and $+$ denote the usual multiplication and addition operations, respectively. Denote the decryption function by D. E is homomorphic with respect to addition, i.e. $D(E(m_1) \times E(m_2)) = D(E(m_1 + m_2))$, which can be easily verified as

$$\begin{aligned} E(m_1) \times E(m_2) &\equiv (g^r, g^{m_1} \times h^r) \times (g^{r'}, g^{m_2} \times h^{r'}) \pmod{p} \\ &\equiv (g^{r+r'}, g^{m_1} \times g^{m_2} \times h^{r+r'}) \pmod{p} \\ &\equiv (g^{r+r'}, g^{m_1+m_2} \times h^{r+r'}) \pmod{p} \\ &\equiv E(m_1 + m_2). \end{aligned}$$

There is no multiplicative homomorphic property of E. In other words, there does not exist a multiplicative operation $\boxtimes$ in the ciphertext space such that

$$D(E(m_1 \times m_2)) = D(E(m_1) \boxtimes E(m_2)).$$

However, E preserves multiplication by an integer scalar. For a random integer scalar k, we have

$$\begin{aligned}
E(m)^k &= ({c_1}^k, {c_2}^k) \\
&\equiv (g^{k\times r}, (g^m \times h^r)^k) \pmod{p} \\
&\equiv (g^{k\times r}, g^{k\times m} \times h^{k\times r}) \pmod{p} \\
&\equiv (g^{r'}, g^{k\times m} \times h^{r'}) \pmod{p} \\
&\equiv E(k\times m)
\end{aligned}$$

with the randomness $r' = k \times r$, which means

$$D(E(m)^k) = D(E(k \times m)).$$

Re-randomization for the additively homomorphic ElGamal algorithm is exactly the same as for the multiplicative version.

5.3 The Elliptic Curve ElGamal Algorithm

The elliptic curve homomorphic ElGamal algorithm, which uses elliptic curve arithmetic over a finite field, is described as an analogue of the additively homomorphic ElGamal algorithm. The group elements in the usual ElGamal algorithm are replaced by points on the elliptic curve, and the multiplication and exponentiation operations are replaced by point addition and point multiplication, respectively. The security of the elliptic curve ElGamal algorithm is based on the Elliptic Curve Discrete Logarithm Problem (ECDLP). (See Sect. 2.4.2.)

5.3.1 Key Generation

For key generation:

(i) Choose a large prime p, generally in the range of $160 \leq |p| \leq 256$.
(ii) Choose an elliptic curve $\mathscr{E}(\mathbb{Z}_p)$ over the finite field of p elements such that the order of the elliptic curve group $\mathscr{E}(\mathbb{Z}_q)$ is divisible by prime q.
(iii) Choose a generating point P of the elliptic curve group $\mathscr{E}$.
(iv) Select a random $d \in \mathbb{Z}_n$ and set $Q = [d]P$, where the operation is point multiplication in $\mathscr{E}$.

The public key is (P, Q, q, p) and the private key is d.

5.3.2 Encryption

The plaintext is $m \in \mathbb{Z}_q$. The sender generates a random number $r \in \mathbb{Z}_q$ and computes the encryption of the plaintext m, which is the ciphertext pair $E(m) = (R_1, R_2)$, where

$$R_1 = [r]P$$
$$R_2 = [m]P \oplus [r]Q,$$

where R_1 and R_2 are points on the given curve.

For encryption of each message, a new r is chosen to ensure the security of the cryptosystem; r should not be guessable.

5.3.3 Decryption

The legitimate receiver holds the private key d and knows P, q and p. So the receiver can decrypt the ciphertext (R_1, R_2) without any knowledge of r, by computing U_1 and U_2 as

$$U_1 = [d]([r]P) = [d \times r]P$$
$$U_2 = -U_1 \oplus R_2 = -[d \times r]P \oplus [m]P \oplus [r \times d]P = [m]P,$$

where $-U_1$ is the additive inverse of U_1 in $\mathscr{E}$. So the owner of the private key obtains $[m]P$ after the decryption process. The decryption requires the solution of an Elliptic Curve Discrete Logarithm Problem (ECDLP) to find m from $[m]P$, which is an intractable problem. However if the message space (or q) is small, then the discrete logarithm can be computed in a reasonable time by trying all the values of $[i]P$ for $i = 1, 2, \ldots, q-1$. In that case the decryption is efficient but the cryptosystem is not secure anymore.

5.3.4 Homomorphic Properties

The encryption function of the elliptic curve ElGamal cryptosystem is described by

$$E : (\mathbb{Z}_q, +) \rightarrow (\mathscr{E} \times \mathscr{E}, \oplus)$$
$$m \mapsto ([r]P, [m]P \oplus [r]Q),$$

where the $+$ operation in the domain is the usual addition, and $\oplus$ in the co-domain is the elliptic curve point addition operation. Let D be the decryption function. E is homomorphic with respect to (usual) addition, i.e. $D(E(m_1) \oplus E(m_2)) = D(E(m_1 + m_2))$, which can be easily verified as

$$\begin{aligned}E(m_1)\oplus E(m_2) &= ([r]P,[m_1]P\oplus[r]Q)\oplus([r']P,[m_2]P\oplus[r']Q)\\ &= ([r]P\oplus[r']P,[m_1]P\oplus[r]Q\oplus[m_2]P\oplus[r']Q)\\ &= ([r+r']P,[m_1+m_2]P\oplus[r']Q)\\ &= E(m_1+m_2)\end{aligned}$$

for randomness $r+r'\in\mathbb{Z}_q$. There is no multiplicative homomorphic property of E. However, E preserves multiplication by an integer scalar. For a random integer scalar $k\in\mathbb{Z}_q$, we have

$$\begin{aligned}[k]E(m) &= [k]([r]P,[m]P\oplus[r]Q)\\ &= ([k\times r]P,[k\times m]P\oplus[k\times r]Q)\\ &= E(k\times m)\end{aligned}$$

for some random $k\times r\in\mathbb{Z}_q$, which means

$$D([k]E(m)) = D(E(k\times m)).$$

Re-randomization is possible for the elliptic curve ElGamal algorithm. Let $E(m)=C=(U_1,U_2)=([r]P,[m]P+[r]Q)$ for random $r\in\mathbb{Z}_q$ and let $r'\in\mathbb{Z}_q$ be another chosen random number. Then

$$\begin{aligned}([r']P,[r']Q)\oplus E(m) &= ([r']P,[r']Q)\oplus(U_1,U_2)\\ &= ([r'+r]P,[m]P\oplus[r'+r]Q),\end{aligned}$$

which is a valid encryption of m since $r'+r\in\mathbb{Z}_q$. Hence

$$D(([r']P,[r']Q)\oplus C) = D(C).$$

Chapter 6
Benaloh Algorithm

The Benaloh algorithm [2], proposed in 1994 by Josh Benaloh, is a generalization of the Goldwasser-Micali (GM) algorithm. While the GM encryption function can encrypt one bit at a time, the Benaloh encryption function can encrypt larger block sizes. The security of the Benaloh algorithm is based on the rth Residuosity Problem (see Sect. 2.1.11), which is intractable when the modulus n is a composite integer with two unknown large prime factors.

6.1 Key Generation

Let r be an odd prime which is the desired block length of the cryptosystem. The plaintext messages are chosen from $\mathbb{Z}_r$. For key generation:

(i) Choose two large prime numbers p and q such that $r|(p-1)$, $\gcd(r,(p-1)/r) = 1$ and $\gcd(r,q-1) = 1$, which yields $r|\phi(n)$ and $\gcd(r,\phi(n)/r) = 1$. Such primes are easy to find and can be selected uniformly in polynomial time; see Lemma 2.27 in [1].
(ii) Set $n = p \times q$ and compute $\phi(n) = (p-1) \times (q-1)$.
(iii) Choose $y \in \mathbb{Z}_n^*$ such that $y^{\phi(n)/r} \not\equiv 1 \pmod{n}$, i.e. y is not an rth residue modulo n by Corollary 4 in Chap. 2.

The public key is (n,y) and the private key is (p,q).

Note that r should be a prime, otherwise decryption of the scheme is incorrect, because it can result in ambiguous decryption of ciphertexts [7].

Ç. K. Koç et al., *Partially Homomorphic Encryption*,
https://doi.org/10.1007/978-3-030-87629-6_6

Algorithm 19 Key generation algorithm for the Benaloh cryptosystem

1: **Input:** Odd prime r. Large primes p and q such that $r|(p-1)$, $\gcd(r, p-1/r) = 1$ and $\gcd(r, q-1) = 1$.
2: **Output:** Public key (n, y) and private key (p, q).
3: $n := p \times q$
4: $\phi := (p-1) \times (q-1)$
5: **repeat**
6: $\quad y :=$ RANDINT$(1, n-1)$
7: **until** GCD(y,n)=1 and LRPOW$(y, \phi/r, n) \neq 1$ ▷ Choose an rth non-residue $y \in \mathbb{Z}_n^*$
8: **return** (n, y) and (p, q)

6.2 Encryption

The plaintext message is $m \in \mathbb{Z}_r$. The sender generates a random number $u \in \mathbb{Z}_n^*$ and computes the ciphertext

$$c \equiv y^m \times u^r \pmod{n}.$$

Algorithm 20 Encryption algorithm for the Benaloh cryptosystem

1: **Input:** Message $m \in \mathbb{Z}_r$ where r is an appropriate odd prime, encryption key (n, y).
2: **Output:** Ciphertext $c \in \mathbb{Z}_n^*$.
3: **repeat**
4: $\quad u :=$ RANDINT$(1, n)$
5: **until** GCD(u,n)=1
6: $c :=$ LRPOW$(y, m, n) \times$ LRPOW$(u, r, n) \pmod{n}$
7: **return** c

6.3 Decryption

Given the encrypted message $c \equiv y^m \times u^r \pmod{n}$, the receiver can easily compute

$$y^{-i} \times c \equiv y^{m-i} \times u^r \pmod{n}$$

for any $i \in \mathbb{Z}_r$. It is easy to observe that $y^{-i} \times c$ is an rth residue if and only if either $m - i = 0$ or y^{m-i} is an rth residue for $0 < m - i < r$. If $m - i$ was the least positive integer, for some $0 < i < r$, making y^{m-i} an rth residue modulo n, then $m - i$ would divide r by Lemma 7, part (i). However, since r is an odd prime, its only proper divisor is 1. So $m - i$ would be 1, which contradicts the fact that y is not an rth residue. Therefore, $y^{-i} \times c$ is an rth residue if and only if $m - i = 0$ (or $m = i$). By Corollary 4, $m = i$ if and only if

$$\begin{aligned}1 &\equiv (y^{-i} \times c)^{\phi(n)/r} \pmod{n}\\ &\equiv (y^{-i} \times y^m \times u^r)^{\phi(n)/r} \pmod{n}\\ &\equiv y^{(m-i)\times\phi(n)/r} \times u^{\phi(n)} \pmod{n}\\ &\equiv y^{(m-i)\times\phi(n)/r} \pmod{n},\end{aligned}$$

where the last equivalence follows from the fact that $u^{\phi(n)} \equiv 1 \pmod{n}$, as $\gcd(u,n) = 1$, by Fermat's Little Theorem (Theorem 8).

Therefore, the receiver who holds the private key (p,q) can decide whether $y^{-i} \times c$ is an rth residue by solving the congruence

$$1 \equiv y^{(m-i)\times\phi(n)/r} \pmod{n}.$$

If r is small, then one can recover m by an exhaustive search, i.e. checking whether $y^{(m-i)\times\phi(n)/r} \equiv 1 \pmod{n}$ for $i = 0,\ldots,r-1$. For larger values of r, by precomputing the values, the baby-step giant-step algorithm can be used to compute the discrete logarithm to recover m.

If the factors p and q of n are unknown, or $\phi(n)$ is unknown, then finding an rth residue is an intractable problem.

Algorithm 21 Decryption algorithm for the Benaloh cryptosystem

1: **Input:** Received message c, appropriate odd prime r, encryption key (n,y) and decryption key (p,q).
2: **Output:** Original message m.
3: $\phi := (p-1)\times(q-1)$
4: **for** $i = 1$ **to** $r-1$ **do**
5: **if** LRPOW((LRPOW($y, -i, n$))$\times c$, ϕ/r, n)$= 1$ **then** $m := i$
6: **return** m

6.4 Homomorphic Properties

The encryption function of the Benaloh cryptosystem is described by

$$\begin{aligned}E : (\mathbb{Z}_r, +) &\rightarrow (\mathbb{Z}_n^*, \times)\\ m &\mapsto y^m \times u^r,\end{aligned}$$

where $+$ and $\times$ denote the usual addition and multiplication operations, respectively. Denote the decryption function by D. If c_1 and c_2 are encryptions of messages m_1 and m_2, then

$$\begin{aligned}E(m_1) = c_1 &\equiv y^{m_1} \times {u_1}^r \pmod{n}\\ E(m_2) = c_2 &\equiv y^{m_2} \times {u_2}^r \pmod{n},\end{aligned}$$

where u_1 and u_2 are randomly selected integers in $\mathbb{Z}_n^*$. Then the homomorphic property is

$$\begin{aligned} E(m_1) \times E(m_2) &\equiv (y^{m_1} \times {u_1}^r) \times (y^{m_2} \times {u_2}^r) \pmod n \\ &\equiv y^{m_1+m_2} \times (u_1 \times u_2)^r \pmod n \\ &\equiv E(m_1 + m_2), \end{aligned}$$

where the randomness in the encryption of $m_1 + m_2$ is $u_1 \times u_2$, which is which is neither uniformly distributed in $\mathbb{Z}_n^*$ nor independent of the randomness in $E(m_1)$ and $E(m_2)$. However, this can be addressed by re-randomization, which is explained at the end of this section.

Then we have

$$D(E(m_1) \times E(m_2)) = D(E(m_1 + m_2)),$$

which verifies that E is additively homomorphic. There is no multiplicative homomorphic property of the encryption function E. In other words, there does not exist a multiplicative operation $\boxtimes$ in the ciphertext space such that

$$D(E(m_1 \times m_2)) = D(E(m_1) \boxtimes E(m_2)).$$

The Benaloh algorithm, however, is multiplicatively and additively homomorphic with an integer scalar k.

For an integer scalar k, we have

$$E(m)^k \equiv y^{k \times m} \times (u^k)^r \pmod n \equiv E(k \times m)$$

with the randomness $u^k \in \mathbb{Z}_n^*$, which means

$$D(E(m)^k) = D(E(k \times m)).$$

Similarly, for an integer scalar k, we have

$$E(m+k) \equiv y^{m+k} \times u^r \pmod n \equiv y^m \times y^k \times u^r \pmod n \equiv E(m) \times y^k \pmod n,$$

which means

$$D(E(m+k)) = D(E(m) \times y^k).$$

Re-randomization is possible for the Benaloh algorithm. Let $u' \in \mathbb{Z}_n^*$ be a random number and $c = E(m)$. Then,

$$(u')^r \times E(m) \equiv (u')^r \times y^m \times u^r \pmod n \equiv y^m \times (u \times u')^r \pmod n,$$

which is a valid encryption of m with randomness $u \times u' \in \mathbb{Z}_n^*$. Hence

$$D((u') \times c) = D(c).$$

6.5 Security

The notion of semantic security has been mentioned in the security part of the Goldwasser-Micali cryptosystem.

Theorem 33 *The Benaloh cryptosystem is semantically secure if and only if the rth Residuosity Problem is intractable.*

Proof. ($\Rightarrow$) Assume that the rth Residuosity Problem is not intractable. Then an adversary can decide the rth residuosity of a given element. Let m_0 and m_1 be two different messages known by the adversary. Let $c \equiv y^m \times u^r \pmod{n}$ be the given encryption of either m_0 or m_1, where u is randomly chosen from $\mathbb{Z}_n^*$. The adversary computes $c \times y^{-m_0} \equiv y^{m-m_0} \times u^r \pmod{n}$. She can decide whether $c \times y^{-m_0}$ is an rth residue. If it is an rth residue, $m - m_0 = 0$, or $m = m_0$, by the arguments used in the decryption process of the cryptosystem. If it is an rth non-residue, then $m = m_1$ with an overwhelming probability. Therefore, the adversary can distinguish the encryption of the given messages, which implies the lack of semantic security.

($\Leftarrow$) Assume that the Benaloh cryptosystem is not semantically secure. Then given a ciphertext c, an adversary can decide whether it is an encryption of 0 or not. If it is an encryption of 0, then c is an rth residue modulo n; otherwise, c is an rth non-residue modulo n. Therefore, the rth Residuosity Problem is not intractable.

6.6 Example

Let $r = 3$, $p = 7$, $q = 5$. Note that r is an odd prime, and $3|(7-1)$ with $\gcd(3,(7-1)/3) = 1$, and $3 \nmid (5-1)$ with $\gcd(3,5-1) = 1$. Compute $n = p \times q = 35$ and $\phi(n) = 24$. Choose $y = 2$. Note that $2 \in \mathbb{Z}_{35}^*$, and $y^{-1} \equiv 2^{-1} \equiv 18 \pmod{35}$, and $y^{\phi(n)/r} \equiv 2^8 \equiv 11 \not\equiv 1 \pmod{35}$. Hence $(n,y) = (35,2)$ is the public encryption key and $(p,q) = (7,5)$ is the private decryption key.

Let $m = 1$ and choose $u = 18$. Compute $c = E(m)$

$$c \equiv y^m \times u^r \equiv 2^1 \times 18^3 \equiv 9 \pmod{35}.$$

Once we get the ciphertext $c = 9$ we compute $y^{-i} \times c$ for $i = 0,1,2$:

$$\begin{aligned} y^0 \times c &\equiv 9 \pmod{35} \\ y^{-1} \times c &\equiv 18 \times 9 \equiv 22 \pmod{35} \\ y^{-2} \times c &\equiv 18^2 \times 9 \equiv 11 \pmod{35} \end{aligned}$$

Finally we check the residuosities of $y^{-i} \times c$ by raising them to the power $\phi(n)/r = 8$:

$$\begin{aligned} 9^8 &\equiv 11 \not\equiv 1 \pmod{35} \\ 22^8 &\equiv 1 \pmod{35} \\ 11^8 &\equiv 16 \not\equiv 1 \pmod{35} \end{aligned}$$

This means that 9 and 11 are 3rd non-residues whereas 22 is a 3rd residue modulo 35, which leads to $m = 1$.

To illustrate the homomorphic property, further let $m' = 2$ with $u' = 17$, then encryption of m' becomes

$$c' \equiv y^{m'} \times (u')^r \equiv 2^2 \times 17^3 \equiv 17 \pmod{35}.$$

Let $c'' = c \times c' = 9 \times 17 \equiv 13 \pmod{35}$. Next we compute $y^{-i} \times c''$ for $i = 0, 1, 2$:

$$\begin{aligned} y^0 \times c'' &\equiv 13 \pmod{35} \\ y^{-1} \times c'' &\equiv 18 \times 13 \equiv 24 \pmod{35} \\ y^{-2} \times c'' &\equiv 18^2 \times 13 \equiv 12 \pmod{35}. \end{aligned}$$

Finally we decrypt c'' by checking the residuosities of $y^{-i} \times c''$ via raising them to the power $\phi(n)/r = 8$ as

$$\begin{aligned} 13^8 &\equiv 1 \pmod{35} \\ 24^8 &\equiv 16 \not\equiv 1 \pmod{35} \\ 12^8 &\equiv 11 \not\equiv 1 \pmod{35}. \end{aligned}$$

This means that 24 and 12 are 3rd non-residues whereas 13 is a 3rd residue modulo 35. Hence $c'' = 13$ decrypts to $m'' = 0$. Note that $m + m' \equiv 1 + 2 \equiv 0 \pmod{3}$, which verifies the equality $D(E(m) \times E(m')) = D(E(m+m')) = m + m'$.

To exemplify the multiplicative homomorphic property with an integer scalar, further let $k = 2$ be a constant. We compute the kth power of the ciphertext $c = 9 = E(1)$ worked before: $c^k \equiv 9^2 \equiv 11 \pmod{35}$, Next we compute $y^{-i} \times c^k$ for $i = 0, 1, 2$:

$$\begin{aligned} y^0 \times c^k &\equiv 11 \pmod{35} \\ y^{-1} \times c^k &\equiv 18 \times 11 \equiv 23 \pmod{35} \\ y^{-2} \times c^k &\equiv 18^2 \times 11 \equiv 29 \pmod{35}. \end{aligned}$$

Finally we decrypt c^k by checking the residuosities of $y^{-i} \times c^k$ via raising them to the power $\phi(n)/r = 8$ as

$$\begin{aligned} 11^8 &\equiv 16 \not\equiv 1 \pmod{35} \\ 23^8 &\equiv 11 \not\equiv 1 \pmod{35} \\ 29^8 &\equiv 1 \pmod{35}. \end{aligned}$$

This means that 11 and 23 are 3rd non-residues whereas 29 is a 3rd residue modulo 35. Hence $c^k = 11$ decrypts to $m^{(k)} = 2$. Note that $k \times m \equiv 2 \times 1 \equiv 2 \pmod 3$, which verifies the equality $D(E(m)^k) = D(E(k \times m)) = k \times m$.

To illustrate the additive homomorphic property with an integer scalar, let $k' = 4$. Further let $c^{(k')} = y^{k'} \times c = 2^4 \times 9 \equiv 4 \pmod{35}$. Next we compute $y^{-i} \times c^{(k')}$ for $i = 0, 1, 2$:

$$\begin{aligned} y^0 \times c^{(k')} &\equiv 4 \pmod{35} \\ y^{-1} \times c^{(k')} &\equiv 18 \times 4 \equiv 2 \pmod{35} \\ y^{-2} \times c^{(k')} &\equiv 18^2 \times 4 \equiv 1 \pmod{35}. \end{aligned}$$

Finally we decrypt $c^{(k')}$ by checking the residuosities of $y^{-i} \times c^{(k')}$ via raising them to the power $\phi(n)/r = 8$ as

$$\begin{aligned} 4^8 &= 16 \not\equiv 1 \pmod{35} \\ 2^8 &\equiv 11 \not\equiv 1 \pmod{35} \\ 1^8 &\equiv 1 \pmod{35}. \end{aligned}$$

This means 2 and 4 are 3rd non-residues whereas 1 is a 3rd residue modulo 35. Hence $c^{(k')} = 4$ decrypts to $m^{(k')} = 2$. Note that $m + k' \equiv 1 + 4 \equiv 2 \pmod 3$, which verifies the equality $D(E(m) \times y^k) = D(E(m+k)) = m+k$.

To illustrate re-randomization, choose a random $u''' = 17$ and compute a re-randomized ciphertext c''' as follows:

$$c''' \equiv c \times (u''')^r \equiv 9 \times 17^3 \equiv 9 \times 13 \equiv 12 \pmod{35}.$$

Next we compute $y^{-i} \times c'''$ for $i = 0, 1, 2$:

$$\begin{aligned} y^0 \times c''' &\equiv 12 \pmod{35} \\ y^{-1} \times c''' &\equiv 18 \times 12 \equiv 6 \pmod{35} \\ y^{-2} \times c''' &\equiv 18^2 \times 12 \equiv 3 \pmod{35}. \end{aligned}$$

Finally we decrypt c''' by checking the residuosities of $y^{-i} \times c'''$ via raising them to the power $\phi(n)/r = 8$ as

$$\begin{aligned} 12^8 &\equiv 11 \not\equiv 1 \pmod{35} \\ 6^8 &\equiv 1 \pmod{35} \\ 3^8 &\equiv 16 \not\equiv 1 \pmod{35} \end{aligned}$$

This means that 12 and 3 are 3rd non-residues whereas 6 is a 3rd residue modulo 35. Hence $c''' = 12$ decrypts to $m''' = m = 1$.

Chapter 7
Naccache-Stern Algorithm

The Naccache-Stern (NS) algorithm [17] proposed by David Naccache and Jacques Stern in 1998 can be viewed as a generalization of the Benaloh algorithm. There are two versions of the NS algorithm, one deterministic and the other probabilistic; the latter one obtained via small modifications of the former one. Both versions are homomorphic encryption schemes. The expansion rate, i.e. the ratio between the lengths of the ciphertext and the plaintext, of the probabilistic version is much better than that of the probabilistic homomorphic encryption schemes proposed so far, namely the Goldwasser-Micali and Benaloh algorithms. The semantic security of the probabilistic NS algorithm is based on the intractability of the Higher Residuosity Problem, (see Sect. 2.1.11), like the security of the Benaloh algorithm.

7.1 The Deterministic Version

We start with the deterministic NS algorithm. As the name suggests, no randomness is involved in encryption process.

7.1.1 Key Generation

For key generation:

(i) Choose k small distinct odd primes $p_1, \ldots, p_k$ (the largest one is 10 bits long), where k is even.
(ii) Set $u = \prod_{i=1}^{k/2} p_i$ and $v = \prod_{i=k/2+1}^{k} p_i$, and $\sigma = u \times v = \prod_{i=1}^{k} p_i > 2^{160}$.
(iii) Choose two large primes a and b such that $p = 2 \times a \times u + 1$ and $q = 2 \times b \times v + 1$ are primes as well. This choice should also satisfy that for $i = 1, \cdots, k/2$,

$$p_i \mid (p-1) \text{ and } \gcd(p_i, \frac{p-1}{p_i}) = 1 \text{ and } \gcd(p_i, q-1) = 1,$$

Ç. K. Koç et al., *Partially Homomorphic Encryption*,
https://doi.org/10.1007/978-3-030-87629-6_7

and for $i = k/2+1,\cdots,k$,

$$p_i \mid (q-1) \text{ and } \gcd(p_i, \frac{q-1}{p_i}) = 1 \text{ and } \gcd(p_i, p-1) = 1.$$

(iv) Set $n = p \times q$ (≥ 768 bits).
The choices in part (iii) require that $\sigma \mid \phi(n)$ and $\gcd(\sigma, \phi(n)/\sigma) = 1$.

(v) Select a semi-random $g \in \mathbb{Z}_n^*$ such that the order of g is $\phi(n)/4$, which ensures that g is not a p_ith residue (or p_ith power) modulo n, i.e. $g^{\phi(n)/p_i} \not\equiv 1 \pmod{n}$, for any $1 \leq i \leq k$, by Corollary 4 in Chap. 2.

The public key is (n, g) and the private key is (p, q). Note that σ can optionally be kept secret.

Furthermore, there is actually no reason why the p_is should be prime. Everything goes through, *mutatis mutandis*, as soon as the p_is are mutually prime. Thus, for example, they can be chosen as prime powers, which is a way to increase the variability of the scheme.

7.1.2 Encryption

The message m is chosen from $\mathbb{Z}_\sigma$. The sender computes the ciphertext

$$c \equiv g^m \pmod{n}.$$

7.1.3 Decryption

First of all, consider a system of congruences $m \equiv m_i \pmod{p_i}$ for $i = 1,\cdots,k$. By the Chinese Remainder Theorem (Theorem 15), m can be calculated modulo $\sigma = p_1 \ldots p_k$.

In order to find m_i, calculate $c_i \equiv c^{\phi(n)/p_i} \pmod{n}$ as

$$\begin{aligned}
c_i &\equiv c^{\phi(n)/p_i} \pmod{n} \\
&\equiv g^{m \times \phi(n)/p_i} \pmod{n} \\
&\equiv g^{(m_i + y_i \times p_i) \times \phi(n)/p_i} \pmod{n} \\
&\equiv g^{m_i \times \phi(n)/p_i} \times g^{y_i \times \phi(n)} \pmod{n} \\
&\equiv g^{m_i \times \phi(n)/p_i} \pmod{n},
\end{aligned}$$

where y_i stands for $(m - m_i)/p_i$ and in the last equivalence $g^{y_i \times \phi(n)}$ vanishes due to $y_i \times \phi(n)$ being a multiple of the order of $\mathbb{Z}_n^*$ from which g is chosen. In order to find the correct value of m_i, the receiver applies exhaustive search over $0 \leq j \leq p_i - 1$

Algorithm 22 Key generation algorithm for the Naccache-Stern Cryptosystem

1: **Input:** Even number k and small odd primes $p_1, \ldots, p_k$.
2: **Output:** Public key (n, g) and private key (p, q).
3: $u := 1$; $v := 1$
4: **for** $i = 1$ **to** $k/2$ **do**
5: $u := u \times p_i$
6: **for** $i = k/2 + 1$ **to** k **do**
7: $v := v \times p_i$
8: $\sigma := u \times v$
9: **procedure**
10: **repeat**
11: $a :=$RANDINT$(2^{309}, 2^{310} - 1)$; $b :=$ RANDINT$(2^{309}, 2^{310} - 1)$
12: **if** ISPRIME (a) is true and ISPRIME (b) is true **then**
13: **if** ISPRIME $(2 \times a \times u + 1)$ is true and ISPRIME $(2 \times b \times v + 1)$ is true **then**
14: **for** $i = 1$ **to** $k/2$ **do**
15: **if** $p_i \mid (2 \times a \times u - 1)$ and GCD$(p_i, (2 \times a \times u - 1)/p_i) = 1$ and GCD$(p_i, 2 \times b \times v - 1) = 1$ **then**
16: **for** $j = k/2 + 1$ **to** k **do**
17: **if** $p_j \mid (2 \times b \times v - 1)$ and GCD$(p_j, (2 \times b \times v - 1)/p_j) = 1$ and GCD$(p_j, 2 \times a \times u - 1) = 1$ **then**
18: **break procedure**
19: $p := 2 \times a \times u + 1$; $q := 2 \times b \times v + 1$
20: $n := p \times q$; $\phi := (p - 1) \times (q - 1)$
21: **procedure**
22: **repeat**
23: $g :=$RANDINT $(2, n - 1)$
24: **until** GCD$(g, n) = 1$
25: $g' := 1$ and $i := 0$
26: **repeat** $g' = g \times g' \pmod n$
27: $i = i + 1$
28: **until** $g' = 1$
29: **if** $i = \phi/4$ **then**
30: **break procedure**
31: **return** (n, g) and (p, q)

Algorithm 23 Encryption algorithm for the Naccache-Stern cryptosystem

1: **Input:** Message $m \in \mathbb{Z}_\sigma$ and encryption key (n, g).
2: **Output:** Ciphertext $c \in \mathbb{Z}_n^*$.
3: c:=LRPOW(g, m, n)
4: **return** c

until $c_i \equiv g^{j \times \phi(n)/p_i} \pmod n$. Once each message m_i is computed, m is recovered by the Chinese Remainder Theorem (Theorem 15).

Algorithm 24 Decryption algorithm for the Naccache-Stern cryptosystem

1: **Input:** Received message c, encryption key (n,g), decryption key (p,q), and even number k of distinct primes $p_1,\ldots,p_k$.
2: **Output:** Original message m.
3: $\phi := (p-1)\times(q-1)$
4: **for** $i=1$ **to** k **do**
5: $\quad c_i := \text{LRPOW}(c, \phi/p_i, n)$
6: $\quad$ **for** $j=0$ **to** $p_i - 1$ **do**
7: $\quad\quad$ **if** $c_i = \text{LRPOW}\ (g, j\times\phi/p_i, n)$ **then** $m_i := j$
8: $x := \text{CRA}(\{m_i\},\{p_i\})$

7.1.4 Security

The security of the deterministic Naccache-Stern cryptosystem is based on the difficulty of both factoring n and taking discrete logarithms. In order to prevent the computation of discrete logarithms by the baby-step giant-step algorithm, σ should be large enough, minimum 2^{160}. Although there is no formal proof that inverting the encryption function is equivalent to factoring, there is no known efficient factoring method to attack the cryptosystem. Actually, the requested size of n (≥ 768 bits) makes factoring n a very hard task. There is a further difficulty when σ is known. For a detailed exploration for this, see [17].

There is no semantic security for this version since it is not probabilistic (or randomized).

7.2 The Probabilistic Version

We now turn to the probabilistic version of the scheme which is a slightly modified version of the deterministic NS algorithm.

7.2.1 Key Generation

Key generation is exactly the same as in the deterministic version. The public key is (n,g,σ) and the private key is (p,q). Note that this version requires σ to be public.

7.2.2 Encryption

The message m is chosen from $\mathbb{Z}_\sigma$. The sender generates a random number $x \in \mathbb{Z}_n$ and computes the ciphertext

$$c \equiv g^m \times x^\sigma \pmod{n}.$$

Algorithm 25 Encryption algorithm for the Naccache-Stern cryptosystem

---

1: **Input:** Message $m \in \mathbb{Z}_\sigma$ and encryption key (n, g).
2: **Output:** Ciphertext $c \in \mathbb{Z}_n^*$.
3: $x :=$ RANDINT$(0, n-1)$
4: c:=LRPOW$(g, m, n)\times$LRPOW$(x, \sigma, n) \pmod{n}$
5: **return** c

7.2.3 Decryption

Although a random factor x is introduced in the encryption step, decryption is the same as in the deterministic version because multiplying by x^σ has no effect during the calculations of the c_i due to $x^{\sigma\times\phi(n)/p_i}$ being 1 modulo n.

7.2.4 Homomorphic Properties

Homomorphic properties of the probabilistic version are presented in this section. However, these properties, except re-randomization, hold for the deterministic version as well.

The encryption function of the probabilistic NS cryptosystem is given by

$$\begin{aligned} E : (\mathbb{Z}_\sigma, +) &\to (\mathbb{Z}_n^*, \times) \\ m &\mapsto g^m \times x^\sigma, \end{aligned}$$

where $+$ and $\times$ denote the usual addition and multiplication operations, respectively. Let D be the decryption function. If c_1 and c_2 are encryptions of messages m_1 and m_2, then

$$\begin{aligned} E(m_1) = c_1 &\equiv g^{m_1} \times {x_1}^\sigma \pmod{n} \\ E(m_2) = c_2 &\equiv g^{m_2} \times {x_2}^\sigma \pmod{n}, \end{aligned}$$

where x_1 and x_2 are randomly selected integers in $\mathbb{Z}_n^*$. Then the homomorphic property is

$$\begin{aligned} E(m_1) \times E(m_2) &\equiv (g^{m_1} \times {x_1}^{\sigma}) \times (g^{m_2} \times {x_2}^{\sigma}) \pmod n \\ &\equiv g^{m_1+m_2} \times (x_1 \times x_2)^{\sigma} \pmod n \\ &\equiv E(m_1 + m_2), \end{aligned}$$

where the randomness in the encryption of $m_1 + m_2$ is $x_1 \times x_2$, which is neither uniformly distributed in $\mathbb{Z}_n^*$ nor independent of the randomness in $E(m_1)$ and $E(m_2)$. However this can be addressed by re-randomization, which is explained at the end of this section.

Then we have

$$D(E(m_1) \times E(m_2)) = D(E(m_1 + m_2)),$$

which verifies that E is homomorphic with respect to addition. The encryption function E has no multiplicative homomorphic property. In other words, there does not exist an operation $\boxtimes$ in the ciphertext space such that

$$D(E(m_1) \boxtimes E(m_2)) = D(E(m_1 \times m_2)).$$

However, E preserves multiplication by an integer scalar. For an integer scalar k, we have

$$E(m)^k \equiv g^{k \times m} \times (x^k)^{\sigma} \pmod n \equiv E(k \times m)$$

with the random number $x^k \in \mathbb{Z}_n^*$, which means

$$D(E(m)^k) = D(E(k \times m)).$$

Another homomorphic property supported by the NS algorithm is addition with an integer scalar. For an integer scalar k, we have

$$\begin{aligned} E(m+k) &\equiv g^{m+k} \times x^{\sigma} \pmod n \\ &\equiv g^m \times g^k \times x^{\sigma} \pmod n \\ &\equiv E(m) \times g^k \pmod n, \end{aligned}$$

which means

$$D(E(m+k)) = D(E(m) \times g^k).$$

Re-randomization is possible for the probabilistic NS algorithm. Let $y \in \mathbb{Z}_n^*$ be a random number and $c = E(m)$. Then,

$$y^{\sigma} \times E(m) \equiv y^{\sigma} \times g^m \times x^{\sigma} \pmod n \equiv g^m \times (x \times y)^{\sigma} \pmod n,$$

which is a valid encryption of m for random $x \times y \in \mathbb{Z}_n^*$. Hence

$$D(y^{\sigma} \times c) = D(c).$$

7.2.5 Security

The notion of semantic security has been mentioned in the security part of the Goldwasser-Micali cryptosystem.

Theorem 34 *If the Higher Residuosity Problem is intractable, then the Naccache-Stern cryptosystem is semantically secure.*

Proof. Assume that the NS cryptosystem is not semantically secure. Then given a ciphertext $c \equiv g^m \times x^\sigma \pmod{n}$, an adversary can decide whether it is an encryption of 0. If it is an encryption of 0, then c is a σth residue modulo n; otherwise, c is a σth non-residue modulo n because g was chosen to be a p_ith non-residue modulo n, which directly implies that g is a σth non-residue modulo n. Therefore, the adversary can decide higher-order residuosity, which makes the Higher Residuosity Problem tractable.

This is proved using a *hybrid technique* in the appendix of the original paper [17].

7.2.6 Example

We use the same key generation given in [17] for $k = 6$. Let $p = 2 \times 101 \times 3 \times 5 \times 7 + 1 = 21211$ and $q = 2 \times 191 \times 11 \times 13 \times 17 + 1 = 928643$ and choose $g = 131$. Here

$$n = 21211 \times 928643 = 19697446673,$$

$p_1 = 3$, $p_2 = 5$, $p_3 = 7$, $p_4 = 11$, $p_5 = 13$, $p_6 = 17$, and

$$\sigma = 3 \times 5 \times 7 \times 11 \times 13 \times 17 = 255255.$$

Hence the public key is $(n, g, \sigma) = (19697446673, 131, 255255)$ and the private key is $(p, q) = (21211, 928643)$. To transform the decryption problem into finding discrete logarithms base $g^{\phi(n)/p_i}$ we need $\phi(n)/p_i$s.

$$\frac{\phi(n)}{p_1} = 6565498940 \qquad \frac{\phi(n)}{p_2} = 3939299364 \qquad \frac{\phi(n)}{p_3} = 2813785260$$

$$\frac{\phi(n)}{p_4} = 1790590620 \qquad \frac{\phi(n)}{p_5} = 1515115140 \qquad \frac{\phi(n)}{p_6} = 1158617460$$

Since powers of $g^{\phi(n)/p_i}$ are very large numbers we use the table for discrete logarithm in [17], which gives the last four digits of $g^{j \times \phi(n)/p_i} \pmod{n}$.

	$i=1$	$i=2$	$i=3$	$i=4$	$i=5$	$i=6$
$j=0$	0001	0001	0001	0001	0001	0001
$j=1$	1966	6544	1967	6273	6043	0372
$j=2$	9560	3339	4968	7876	4792	7757
$j=3$		9400	1765	8720	0262	3397
$j=4$		5479	6701	7994	0136	0702
$j=5$			6488	8651	6291	4586
$j=6$			2782	4691	0677	8135
$j=7$				9489	1890	3902
$j=8$				8537	6878	5930
$j=9$				2312	2571	6399
$j=10$				7707	7180	6592
$j=11$					8291	9771
$j=12$					0678	0609
$j=13$						7337
$j=14$						6892
$j=15$						3370
$j=16$						3489

The table above helps us to determine j_is such that $m \equiv j_i \pmod{p_i}$. After calculating the j_is we need to use the Chinese Remainder Theorem to recover $m \in \mathbb{Z}_\sigma$. This involves calculating $M_i = \sigma/p_i$ and $C_i \equiv M_i^{-1} \pmod{p_i}$.

	$i=1$	$i=2$	$i=3$	$i=4$	$i=5$	$i=6$
M_i	85085	51051	36465	23205	19635	15015
C_i	2	1	4	2	8	13

Let $m = 12$, $x = 2$. Then

$$\begin{aligned} c &\equiv g^m \times x^\sigma \pmod{n} \\ &\equiv 131^{12} \times 2^{255255} \pmod{19697446673} \\ &\equiv 13999130218 \times 5499887886 \pmod{19697446673} \\ &\equiv 8320559922 \pmod{19697446673} \end{aligned}$$

is the encrypted message. For decryption we need to calculate $c^{\phi(n)/p_i} \pmod{n}$, $i = 1, \cdots, 6$.

$$
\begin{aligned}
c^{\Phi(n)/p_1} &\equiv 8320559922^{6565498940} \equiv 1 \pmod{19697446673},\\
c^{\Phi(n)/p_2} &\equiv 8320559922^{3939299364} \equiv 339883339 \pmod{19697446673},\\
c^{\Phi(n)/p_3} &\equiv 8320559922^{2813785260} \equiv 19416996488 \pmod{19697446673},\\
c^{\Phi(n)/p_4} &\equiv 8320559922^{1790590620} \equiv 16339936273 \pmod{19697446673},\\
c^{\Phi(n)/p_5} &\equiv 8320559922^{1515115140} \equiv 13791540678 \pmod{19697446673},\\
c^{\Phi(n)/p_6} &\equiv 8320559922^{1158617460} \equiv 418450609 \pmod{19697446673}.
\end{aligned}
$$

By looking at the table of $g^{\Phi(n)/p_i}$s we get

$$
\begin{aligned}
m &\equiv r_1 \pmod{p_1} \equiv 0 \pmod{3},\\
m &\equiv r_2 \pmod{p_2} \equiv 2 \pmod{5},\\
m &\equiv r_3 \pmod{p_3} \equiv 5 \pmod{7},\\
m &\equiv r_4 \pmod{p_4} \equiv 1 \pmod{11},\\
m &\equiv r_5 \pmod{p_5} \equiv 12 \pmod{13},\\
m &\equiv r_6 \pmod{p_6} \equiv 12 \pmod{17},
\end{aligned}
$$

where the original message m is recovered as follows by the Chinese Remainder Theorem.

$$
\begin{aligned}
m &\equiv r_1 \times C_1 \times M_1 + \cdots + r_6 \times C_6 \times M_6 \pmod{\sigma}\\
&\equiv 0 \times 85085 \times 2 + 2 \times 51051 \times 1 + 5 \times 36465 \times 4\\
&\quad + 1 \times 23205 \times 2 + 12 \times 19635 \times 8 + 12 \times 15015 \times 13 \pmod{255255}\\
&\equiv 12 \pmod{255255}.
\end{aligned}
$$

To illustrate the homomorphic property, let $m_1 = 12$ with $x_1 = 2$ and $m_2 = 63$ with $x_2 = 3$. By previous calculations we know that $c_1 = E(m) = 8320559922$.

$$
\begin{aligned}
c_2 &\equiv g^{m_2} \times x_2^{\sigma} \pmod{n}\\
&\equiv 131^{63} \times 3^{255255} \pmod{19697446673}\\
&\equiv 5607754676 \times 7615151865 \pmod{19697446673}\\
&\equiv 781742477 \pmod{19697446673}.
\end{aligned}
$$

Now assume that the encrypted message is

$$
c' \equiv c_1 \times c_2 \equiv 8320559922 \times 781742477 \equiv 3233025890 \pmod{19697446673}.
$$

$$(c')^{\phi(n)/p_1} \equiv 3233025890^{6565498940} \equiv 1 \pmod{19697446673},$$
$$(c')^{\phi(n)/p_2} \equiv 3233025890^{3939299364} \equiv 1 \pmod{19697446673},$$
$$(c')^{\phi(n)/p_3} \equiv 3233025890^{2813785260} \equiv 19416996488 \pmod{19697446673},$$
$$(c')^{\phi(n)/p_4} \equiv 3233025890^{1790590620} \equiv 5941222312 \pmod{19697446673},$$
$$(c')^{\phi(n)/p_5} \equiv 3233025890^{1515115140} \equiv 6691837180 \pmod{19697446673},$$
$$(c')^{\phi(n)/p_6} \equiv 3233025890^{1158617460} \equiv 6931563902 \pmod{19697446673}.$$

By looking at the table of $g^{\phi(n)/p_i}$s we get

$$m' \equiv r'_1 \pmod{p_1} \equiv 0 \pmod 3,$$
$$m' \equiv r'_2 \pmod{p_2} \equiv 0 \pmod 5,$$
$$m' \equiv r'_3 \pmod{p_3} \equiv 5 \pmod 7,$$
$$m' \equiv r'_4 \pmod{p_4} \equiv 9 \pmod{11},$$
$$m' \equiv r'_5 \pmod{p_5} \equiv 10 \pmod{13},$$
$$m' \equiv r'_6 \pmod{p_6} \equiv 7 \pmod{17},$$

where the original message m' is recovered as follows by the Chinese Remainder Theorem.

$$\begin{aligned} m' &\equiv r'_1 \times C_1 \times M_1 + \cdots + r'_6 \times C_6 \times M_6 \pmod{\sigma} \\ &\equiv 0 \times 85085 \times 2 + 0 \times 51051 \times 1 + 5 \times 36465 \times 4 \\ &\quad + 9 \times 23205 \times 2 + 10 \times 19635 \times 8 + 7 \times 15015 \times 13 \pmod{255255} \\ &\equiv 75 \pmod{255255}. \end{aligned}$$

On the other hand,

$$m_1 + m_2 \pmod{\sigma} \equiv 12 + 63 \equiv 75 \pmod{255255},$$

which verifies the homomorphic property $D(E(m_1) \times E(m_2)) = m_1 + m_2$.

To exemplify the multiplicative homomorphic property with an integer scalar, let $m = 12$, $x = 2$, and let $k = 3$ be a scalar. We already have $c = E(m) = 8320559922$. Consider the encrypted message

$$c^k \equiv 8320559922^3 \equiv 2301536931 \pmod{19697446673}.$$

$$(c^k)^{\phi(n)/p_1} \equiv 2301536931^{6565498940} \equiv 1 \pmod{19697446673},$$
$$(c^k)^{\phi(n)/p_2} \equiv 2301536931^{3939299364} \equiv 4922736544 \pmod{19697446673},$$
$$(c^k)^{\phi(n)/p_3} \equiv 2301536931^{2813785260} \equiv 2564911967 \pmod{19697446673},$$
$$(c^k)^{\phi(n)/p_4} \equiv 2301536931^{1790590620} \equiv 11191538720 \pmod{19697446673},$$

$$(c^k)^{\phi(n)/p_5} \equiv 2301536931^{1515115140} \equiv 6691837180 \pmod{19697446673},$$
$$(c^k)^{\phi(n)/p_6} \equiv 2301536931^{1158617460} \equiv 19441917757 \pmod{19697446673}.$$

By looking at the table of $g^{\phi(n)/p_i}$s we get

$$\begin{aligned}
m^{(k)} &\equiv r_1^{(k)} \pmod{p_1} \equiv 0 \pmod{3},\\
m^{(k)} &\equiv r_2^{(k)} \pmod{p_2} \equiv 1 \pmod{5},\\
m^{(k)} &\equiv r_3^{(k)} \pmod{p_3} \equiv 1 \pmod{7},\\
m^{(k)} &\equiv r_4^{(k)} \pmod{p_4} \equiv 3 \pmod{11},\\
m^{(k)} &\equiv r_5^{(k)} \pmod{p_5} \equiv 10 \pmod{13},\\
m^{(k)} &\equiv r_6^{(k)} \pmod{p_6} \equiv 2 \pmod{17},
\end{aligned}$$

where the original message $m^{(k)}$ is recovered as follows by the Chinese Remainder Theorem.

$$\begin{aligned}
m^{(k)} &\equiv r_1^{(k)} \times C_1 \times M_1 + \cdots + r_6^{(k)} \times C_6 \times M_6 \pmod{\sigma}\\
&\equiv 0 \times 85085 \times 2 + 1 \times 51051 \times 1 + 1 \times 36465 \times 4\\
&\quad +3 \times 23205 \times 2 + 10 \times 19635 \times 8 + 2 \times 15015 \times 13 \pmod{255255}\\
&\equiv 36 \pmod{255255}.
\end{aligned}$$

On the other hand, $m \times k \equiv 12 \times 3 \equiv 36 \equiv m^{(k)} \pmod{255255}$. This verifies the equality $D(c^k) = D(E(k \times m)) = k \times m$.

To illustrate the additive homomorphic property with an integer scalar, let $k' = 63$ be a scalar. Consider $m + k'$ as a message. Then

$$\begin{aligned}
E(m+k') &\equiv g^{m+k'} \times x^{\sigma} \pmod{n}\\
&\equiv 131^{75} \times 2^{255255} \pmod{19697446673}\\
&\equiv 3110748894 \times 5499887886 \pmod{19697446673}\\
&\equiv 2428576993 \pmod{19697446673}.
\end{aligned}$$

We know that $c = E(m) = 8320559922$. Furthermore,

$$g^{k'} \times E(m) \equiv 131^{63} \times 8320559922 \equiv 2428576993 \pmod{19697446673},$$

which verifies the equality $g^k \times E(m) = E(m+k)$.

To test re-randomization, again consider $m = 12$, $x = 2$. Then $c = E(m) = 8320559922$ by previous calculations. Let $y = 3$ be a random element of $\mathbb{Z}_n^*$ and calculate

$$c'' \equiv y^{\sigma} \times E(m) \equiv 3^{255255} \times 8320559922 \equiv 13334122605 \pmod{19697446673}.$$

To decrypt c'' we need to calculate $(c'')^{\phi(n)/p_i} \pmod{n}$, $i = 1, \cdots, 6$.

$$\begin{aligned}
(c'')^{\phi(n)/p_1} &\equiv 13334122605^{6565498940} \equiv 1 \pmod{19697446673},\\
(c'')^{\phi(n)/p_2} &\equiv 13334122605^{3939299364} \equiv 339883339 \pmod{19697446673},\\
(c'')^{\phi(n)/p_3} &\equiv 13334122605^{2813785260} \equiv 19416996488 \pmod{19697446673},\\
(c'')^{\phi(n)/p_4} &\equiv 13334122605^{1790590620} \equiv 16339936273 \pmod{19697446673},\\
(c'')^{\phi(n)/p_5} &\equiv 13334122605^{1515115140} \equiv 13791540678 \pmod{19697446673},\\
(c'')^{\phi(n)/p_6} &\equiv 13334122605^{1158617460} \equiv 418450609 \pmod{19697446673}.
\end{aligned}$$

By looking at the table of $g^{\phi(n)/p_i}$s we get

$$\begin{aligned}
m'' &\equiv r_1'' \pmod{p_1} \equiv 0 \pmod{3},\\
m'' &\equiv r_2'' \pmod{p_2} \equiv 2 \pmod{5},\\
m'' &\equiv r_3'' \pmod{p_3} \equiv 5 \pmod{7},\\
m'' &\equiv r_4'' \pmod{p_4} \equiv 1 \pmod{11},\\
m'' &\equiv r_5'' \pmod{p_5} \equiv 12 \pmod{13},\\
m'' &\equiv r_6'' \pmod{p_6} \equiv 12 \pmod{17},
\end{aligned}$$

where the original message m'' is recovered as follows by the Chinese Remainder Theorem.

$$\begin{aligned}
m'' &\equiv r_1'' \times C_1 \times M_1 + \cdots + r_6'' \times C_6 \times M_6 \pmod{\sigma}\\
&\equiv 0 \times 85085 \times 2 + 2 \times 51051 \times 1 + 5 \times 36465 \times 4\\
&\quad + 1 \times 23205 \times 2 + 12 \times 19635 \times 8 + 12 \times 15015 \times 13 \pmod{255255}\\
&\equiv 12 \pmod{255255}.
\end{aligned}$$

Therefore $m'' = m$, which means $y^{\sigma} \times E(m)$ is a valid encryption of m.

Chapter 8
Okamoto-Uchiyama Algorithm

The Okamoto-Uchiyama algorithm was introduced by Tatsuaki Okamoto and Shigenori Uchiyama in 1998 [18]. The security of this public-key cryptosystem is based on the Integer Factorization Problem where $n = p^2 \times q$ (see Sect. 2.1.1), where p and q are k-bit primes, and the p-Subgroup Problem (PSUB) (see Sect. 2.2.9). Both problems are considered as intractable.

8.1 Key Generation

Assume that m is an at most k-bit message. Key generation involves three steps.

(i) Choose distinct large k-bit primes p and q such that $\gcd(p, q-1) = 1$ and $\gcd(p-1, q) = 1$, and compute $n = p^2 \times q$ (which is $3k$ bits long).
(ii) Choose $g \in \mathbb{Z}_n^*$ such that $\text{ord}(g^{p-1})$ is p in $\mathbb{Z}_{p^2}^*$.
(iii) Set $h \equiv g^n \pmod{n}$.

The quadruple (n, g, k, h) is released as the public encryption key but the private decryption key (p, q) is kept secret. h is a supplementary parameter to improve the efficiency of encryption, since it can be calculated easily from g and n.

Algorithm 26 Key generation for the Okamoto-Uchiyama cryptosystem

```
1: Input: Large odd distinct k-bit primes p,q such that gcd(p,q−1) = 1 and gcd(q,p−1) = 1.
2: Output: Public key (n,g,h) and secret key (p,q).
3: n := p^2 × q
4: repeat
5:     g:=RANDINT(1,n−1)                    ▷ Chooses encryption base.
6:     g':=LRPOW(g,p−1,p^2)
7:     g'':=LRPOW(g,p^2−p,p^2)              ▷ Calculates (g^(p−1))^p modulo p^2.
8: until GCD(g,n)=1 and g' ≠ 1 and g'' = 1
9:     h:=LRPOW(g,n,n)
10:    return (n,g,h) and (p,q)
```


Ç. K. Koç et al., *Partially Homomorphic Encryption*,
https://doi.org/10.1007/978-3-030-87629-6_8

8.2 Encryption

After converting an alphanumeric plaintext message M into a purely numeric message m, where $0 \leq m < p$, the sender chooses a random $r \in \mathbb{Z}_n$. Then the encrypted message c is computed by

$$c \equiv g^m \times h^r \pmod{n}. \tag{8.1}$$

A different r-value is assigned to each message in order to ensure security.

Algorithm 27 Encryption algorithm for the Okamoto-Uchiyama cryptosystem

1: **Input:** Message m, encryption key (n, g, h).
2: **Output:** Encrypted message c.
3: r:=RANDINT$(0, n-1)$
4: c:=LRPOW$(g, m, n) \times$ LRPOW$(h, r, n) \pmod{n}$
5: **return** c

8.3 Decryption

Once the receiver gets the encrypted message c, the original message m can be recovered using (p, q) as follows:

$$m \equiv L(c^{p-1} \pmod{p^2}) \times (L(g^{p-1} \pmod{p^2}))^{-1} \pmod{p}\,, \tag{8.2}$$

where $L(z) = (z-1)/p$ is called the logarithm function. The reason we refer to L in that manner is that L is used to extract the exponent of g in $g^m \times h^r$ modulo n and L operates similarly to the usual logarithm function defined on real numbers (see Lemma 13). The rest of this section is devoted to discussions on why Eq. (8.2) works.

Choice of Base g and Well-Definedness of Logarithm Function

Since $\left|\mathbb{Z}_{p^2}^*\right| = p \times (p-1)$, there exists a Sylow p-subgroup of order p of $\mathbb{Z}_{p^2}^*$ by Theorem 25.

Proposition 4 *The set* $H = \{1, p+1, 2 \times p+1, \ldots, (p-1) \times p+1\}$ *is a Sylow p-subgroup of* $\mathbb{Z}_{p^2}^*$.

Proof. Any Sylow p-subgroup of $\mathbb{Z}_{p^2}^*$ has order p, because the largest p power factor that divides $p \times (p-1)$ is p. H contains exactly p elements and it is generated by $p+1$, since

$$(p+1)^{\nu} = 1 + \binom{\nu}{1} \times p + \mathcal{O}(p^2) \equiv 1 + \nu \times p \pmod{p^2},$$

where $\mathcal{O}(p^2)$ denotes the sum of the terms with larger exponents of p. Hence H is a Sylow p-subgroup of $\mathbb{Z}^*_{p^2}$.

Indeed H is the unique Sylow p-subgroup of $\mathbb{Z}^*_{p^2}$ by Corollary 6. Let $L : H \to \mathbb{Z}_p$ be the logarithm function defined by $L(z) = (z-1)/p$. L is well-defined over H, since $z \in H$ implies $z = 1 + \nu \times p$ with $0 \leq \nu < p$. Hence

$$L(z) = \frac{z-1}{p} = \frac{\nu \times p}{p} = \nu \in \mathbb{Z}_p.$$

Recall that the decryption process involves finding the inverse of $L(g^{p-1} \pmod{p^2})$ modulo p, which requires this logarithm not to be 0. Since we have chosen $g \in \mathbb{Z}^*_n$ such that g^{p-1} has order p in $\mathbb{Z}^*_{p^2}$, and since H is the unique p-Sylow subgroup of $\mathbb{Z}^*_{p^2}$, then necessarily we have $g^{p-1} \in H$, i.e. $g^{p-1} \equiv 1 + \nu \times p \pmod{p^2}$ for some $0 < \nu < p$. Here $\nu \neq 0$, otherwise $g^{p-1} \equiv 1 \pmod{p^2}$, which is impossible since g^{p-1} has order p modulo p^2. It turns out that choosing g^{p-1} from H ensures the well-definedness of L.

Finally we prove that L is an isomorphism, which means it determines the exponent of $g^{p-1} \in \mathbb{Z}^*_n$ in a unique way for any $g^{p-1} \in H$.

Lemma 13 *For $a, b \in H$,*

$$L(a \times b \pmod{p^2}) \equiv L(a) + L(b) \pmod{p}, \tag{8.3}$$

and L is an isomorphism.

Proof. Let $a, b \in H$. Since H is a subgroup of $(\mathbb{Z}^*_{p^2}, \times)$ it is closed under the group operation, which is multiplication modulo p^2. Therefore $a \times b \in H$ and $L(a \times b \pmod{p^2})$ is defined. First we transform $a \times b - 1$ in the following manner:

$$a \times b - 1 = (a-1) \times (b-1) + (a-1) + (b-1).$$

In this case,

$$\begin{aligned}
L(a \times b \pmod{p^2}) &\equiv \frac{a \times b - 1}{p} \pmod{p} \\
&\equiv \frac{(a-1) \times (b-1) + (a-1) + (b-1)}{p} \pmod{p} \\
&\equiv (a-1) \times \frac{b-1}{p} + \frac{a-1}{p} + \frac{b-1}{p} \pmod{p} \\
&\equiv (a-1) \times L(b) + L(a) + L(b) \pmod{p} \\
&\equiv L(a) + L(b) \pmod{p},
\end{aligned}$$

since $a \in H$ and $a - 1 \equiv 0 \pmod{p}$ by the definition of H. This property of L implies that L is a group homomorphism. To prove that it is bijective, it is enough to show that L is one-to-one, since $|H| = p = |\mathbb{Z}_p|$. Suppose that $a, b \in H$ and that $L(a) = L(b)$. So

$$\begin{aligned} \frac{a-1}{p} \equiv \frac{b-1}{p} &\pmod{p} \\ a - 1 \equiv b - 1 &\pmod{p^2} \\ a \equiv b &\pmod{p^2}, \end{aligned}$$

which means $a = b$, since a and b are congruence classes modulo p^2. So L is bijective and hence it is an isomorphism.

Algorithm 28 Algorithm for discrete logarithm of the Okamoto-Uchiyama cryptosystem

```
1: function DLFOU(c, p)
2:     Input: Integer c, prime p taken from the decryption key.
3:     Output: Logarithm of c.
4:     z:=LRPOW(c, p − 1, p^2)
5:     d:=(z − 1)/p
6:     return d
```

Correctness of Decryption Function

With the help of the information we have so far, we can prove the correctness of the decryption algorithm.

Theorem 35 *Equivalence (8.2) is correct.*

Proof. Suppose that $n = p^2 \times q$. Let m be a message and c be the encryption of m with randomness r and base g.

$$\begin{aligned} L(c^{p-1} \ (mod\ p^2)) &\equiv L((g^m \times h^r)^{p-1} \pmod{p^2}) \pmod{p} \\ &\equiv L((g^m \times g^{n \times r})^{p-1} \pmod{p^2}) \pmod{p} \\ &\equiv L((g^{p-1})^m \times g^{(p-1) \times n \times r} \pmod{p^2}) \pmod{p} \\ &\equiv L((g^{p-1})^m \times g^{(p-1) \times p^2 \times q \times r} \pmod{p^2}) \pmod{p} \\ &\equiv L((g^{p-1})^m \pmod{p^2}) \pmod{p}, \end{aligned}$$

where the last equivalence follows from the fact that $g \in \mathbb{Z}_{p^2}^*$ and $\mathrm{ord}(\mathbb{Z}_{p^2}) = p \times (p-1)$.

$$
\begin{aligned}
&L((g^{p-1})^m \pmod{p^2}) \\
&\equiv L(g^{p-1} \times g^{p-1} \cdots g^{p-1} \pmod{p^2}) \pmod{p} \\
&\equiv L(g^{p-1} \pmod{p^2}) + \cdots + L(g^{p-1} \pmod{p^2}) \pmod{p} \\
&\equiv m \times L(g^{p-1} \pmod{p^2}) \pmod{p}
\end{aligned}
$$

where multiplication and addition in the above equations are repeated m times. For short, let $L(g^{p-1} \pmod{p^2}) \equiv \mu \pmod{p}$. Then the decryption of $c = g^m \times h^r$ becomes

$$
\begin{aligned}
&L(c^{p-1} \pmod{p^2}) \times (L(g^{p-1} \pmod{p^2}))^{-1} \\
&\equiv L(c^{p-1} \pmod{p^2}) \times \mu^{-1} \pmod{p} \\
&\equiv m \times L(g^{p-1} \pmod{p^2}) \times \mu^{-1} \pmod{p} \\
&\equiv m \times \mu \times \mu^{-1} \pmod{p} \\
&\equiv m \pmod{p},
\end{aligned}
$$

and this completes the proof.

Algorithm 29 Decryption algorithm for the Okamoto-Uchiyama cryptosystem

1: **Input:** Received message c, base of encryption g, decryption key p.
2: **Output:** Original message m.
3: m:=DLFOU(c)×INVERSE(DLFOU(g),p) $\pmod{p}$
4: **return** m

8.4 Homomorphic Properties

Let $n = p^2 \times q$ and $g \in \mathbb{Z}_n^*$ such that $\mathrm{ord}(g^{p-1})$ is p in $\mathbb{Z}_{p^2}^*$. Then the encryption function can be described as:

$$
\begin{aligned}
E_g : \mathbb{Z}_p \times \mathbb{Z}_n &\to \mathbb{Z}_n^* \\
(m, r) &\mapsto g^m \times h^r,
\end{aligned}
$$

where $r \in \mathbb{Z}_n$ is random. Recall that the sender changes the randomness $r \in \mathbb{Z}_n$ for every message to be encrypted. Denote the decryption function by D. Consider the messages m_1 and m_2 with randomness r_1 and r_2, respectively. Then the Okamoto-Uchiyama algorithm is considered to be additively homomorphic as follows:

$$
\begin{aligned}
E_g(m_1, r_1) \times E_g(m_2, r_2) &\equiv (g^{m_1} \times h^{r_1}) \times (g^{m_2} \times h^{r_2}) \pmod{n} \\
&\equiv g^{m_1+m_2} \times h^{r_1+r_2} \pmod{n} \\
&\equiv E_g(m_1 + m_2, r_1 + r_2),
\end{aligned}
$$

where the randomness of the message $m_1 + m_2$ is $r_1 + r_2$, which is neither uniformly distributed in $\mathbb{Z}_n^*$ nor independent of the randomness in $E_g(m_1, r_1)$ and $E_g(m_2, r_2)$. However this can be addressed by re-randomization, which is explained at the end of this section. This means that if we denote $E_g(m, r) = E(m)$, then for two messages m_1 and m_2 we have

$$D(E(m_1 + m_2)) = D(E(m_1) \times E(m_2)). \tag{8.4}$$

On the other hand, no multiplicatively homomorphic operation $\boxtimes$ is known such that

$$D(E(m_1 \times m_2)) = D(E(m_1) \boxtimes E(m_2))$$

for the Okamoto-Uchiyama algorithm. So the Okamoto-Uchiyama encryption scheme is partially homomorphic.

For a message m with randomness r and a scalar k, this algorithm preserves scalar multiplication by k.

$$\begin{aligned} (E_g(m,r))^k &\equiv (g^m \times h^r)^k \pmod{n} \\ &\equiv g^{k \times m} \times h^{k \times r} \pmod{n} \\ &= E_g(k \times m, k \times r), \end{aligned}$$

which means

$$D(E(k \times m)) = D((E(m))^k). \tag{8.5}$$

Another homomorphic property supported by the Okamoto-Uchiyama algorithm is addition with a constant. Let $c = E_g(m, r)$. Then

$$\begin{aligned} E_g(m+k, r) &\equiv g^{m+k} \times (h)^r \pmod{n} \\ &\equiv (g^m \times g^k) \times h^r \pmod{n} \\ &\equiv (g^m \times h^r) \times g^k \pmod{n} \\ &\equiv c \times g^k \pmod{n}, \end{aligned}$$

which means

$$D(E(k+m)) = D(E(m) \times g^k). \tag{8.6}$$

Re-randomization of messages is also possible for the Okamoto-Uchiyama algorithm. Let $r' \in \mathbb{Z}_n$. Calculating

$$\begin{aligned} c \times h^{r'} &\equiv E_g(m,r) \times h^{r'} \pmod{n} \\ &\equiv g^m \times h^r \times h^{r'} \pmod{n} \\ &\equiv g^m \times h^{r+r'} \pmod{n} \\ &= E_g(m, r+r'), \end{aligned}$$

we see that re-randomization gives us another valid encryption of message m, since $r + r' \in \mathbb{Z}_n$. In other words,

$$D(c) = D(c \times h^{r'}). \tag{8.7}$$

8.5 Security

The security of the Okamoto-Uchiyama algorithm relies on the Integer Factorization Problem (Fact[n]).

Theorem 36 *Inverting the Okamoto-Uchiyama encryption function without knowing (p,q) is intractable if and only if Fact$[n]$ for $n = p^2 \times q$ is intractable.*

Proof. ($\Rightarrow$) The secret decryption key being (p,q), it is obvious that knowledge of the factorization of $n = p^2 \times q$ is sufficient to decrypt a ciphertext c.

($\Leftarrow$) Conversely, decrypting $c \equiv g^m \times h^r \equiv g^{m+n\times r} \pmod{n}$ without knowing the factorization of n is equivalent to factoring n. The adversary should solve the equation $c \equiv g^{\alpha} \pmod{n}$ for some $\alpha \in \mathbb{Z}_n$. With a very high probability $p \leq \alpha$, which means $\alpha \notin \mathbb{Z}_p$. So the adversary can decrypt some other correct message $m' \in \mathbb{Z}_p$ such that $\alpha \equiv m' \pmod{p}$ and $m \not\equiv m' \pmod{n}$ with a non-negligible probability. Therefore the adversary can calculate $\gcd(m' - m, n)$ and this gives a nontrivial factor of n. Hence factorization of n follows.

The semantic security of the Okamoto-Uchiyama algorithm relies on the intractability of PSUB.

Theorem 37 *The Okamoto-Uchiyama algorithm is semantically secure if and only if PSUB is intractable.*

Proof. First note that cracking PSUB is the same as distinguishing encryptions of 0 and 1. Calculating

$$\begin{aligned}
c_0^{(p-1)} &\equiv (E_g(0, r_0))^{(p-1)} \pmod{p^2} \\
&\equiv (g^0 \times h^{r_0})^{(p-1)} \pmod{p^2} \\
&\equiv (g^{n\times r_0})^{(p-1)} \pmod{p^2} \\
&\equiv g^{n\times r_0 \times (p-1)} \pmod{p^2} \\
&\equiv g^{p^2\times (p-1)\times q\times r_0} \pmod{p^2} \\
&\equiv 1 \pmod{p^2}
\end{aligned}$$

and

$$\begin{aligned} c_1^{(p-1)} &\equiv (E_g(1,r_1))^{(p-1)} \pmod{p^2} \\ &\equiv (g^1 \times h^{r_1})^{(p-1)} \pmod{p^2} \\ &\equiv (g \times g^{n \times r_1})^{(p-1)} \pmod{p^2} \\ &\equiv g^{(p-1)} \times g^{n \times r_1 \times (p-1)} \pmod{p^2} \\ &\equiv g^{(p-1)} \times g^{p^2 \times (p-1) \times q \times r_1} \pmod{p^2} \\ &\equiv g^{(p-1)} \pmod{p^2}. \end{aligned}$$

Note that $\text{ord}_{\mathbb{Z}_{p^2}}(c_0^{p-1})$ divides $p-1$ and $\text{ord}_{\mathbb{Z}_{p^2}}(c_1^{p-1}) = p$, since $\text{ord}_{\mathbb{Z}_{p^2}}(g^{p-1}) = p$. Thus c_1^{p-1} is in the unique p-subgroup of $\mathbb{Z}_n$ and c_0^{p-1} is not.

($\Rightarrow$) Assume that PSUB is not intractable. Then an adversary A can distinguish between c_0 and c_1 with overwhelming probability. Let m_0 and m_1 be two messages such that $m_0 < m_1$. We will show that with the help of c_0 and c_1, given an encrypted message c, which is known to be the encryption of either m_0 or m_1, A can decide whether it is the encryption of m_0 or m_1. Assume that $c \equiv E_g(m,r)$, where $r \in \mathbb{Z}_n$ is random. A calculates

$$\begin{aligned} x &\equiv c \times (g^{m_0})^{-1} \pmod{n} \\ &\equiv g^m \times h^r \times (g^{m_0})^{-1} \pmod{n} \\ &\equiv g^{(m-m_0)+r \times n} \pmod{n}, \end{aligned}$$

and sets $N = \text{lcm}(p-1, q-1)$, $g' \equiv g^{(m_1-m_0)+r' \times n} \pmod{n}$ and $h' \equiv (g')^n \pmod{n}$, where $r' \in \mathbb{Z}_n$ is random. When $c \equiv E_g(m_1, r) \pmod{n}$,

$$\begin{aligned} x &\equiv g^{(m_1-m_0)+r \times n} \pmod{n} \\ &\equiv g' \times (h')^{t_1} \pmod{n} \\ &\equiv E_{g'}(1, t_1) \pmod{n}, \end{aligned}$$

where $t_1 \equiv (r - r') \times (m_1 - m_0 + r' \times n)^{-1} \pmod{N}$. Here t_1 is defined with non-negligable probability. When $c \equiv E_g(m_0, r) \pmod{n}$,

$$\begin{aligned} x &\equiv g^{(m_0-m_0)+r \times n} \pmod{n} \\ &\equiv (h')^{t_0} \pmod{n} \\ &\equiv E_{g'}(0, t_0) \pmod{n}, \end{aligned}$$

where $t_0 \equiv r \times (m_1 - m_0 + r' \times n)^{-1} \pmod{N}$. Here t_0 is defined with non-negligable probability. So A calculates x, which turns out to be a valid encryption of 0 or 1. If x is a valid encryption of 1, then c is $E_g(m_1, r)$, and if x is a valid encryption of 0, then c is $E_g(m_0, r)$. Since A can distinguish between encryptions of 0 and 1, she can distinguish encryptions of m_1 and m_0. Therefore the Okamoto-Uchiyama algorithm is not semantically secure if PSUB is not intractable.

($\Leftarrow$) Conversely, assume that the Okamoto-Uchiyama algorithm is not semantically secure. In other words, there is an adversary A who can distinguish encryptions of two given messages m_1 and m_0 ($m_0 < m_1$) with non-negligable probability. We will prove that A can also distinguish encryptions of 0 and 1. Given a ciphertext $c \equiv E_g(m,r) \pmod n$, which is either $E_g(1,r_1)$ or $E_g(0,r_0)$, A calculates $c' \equiv c^{m_1-m_0} \times g^{m_0} \pmod n$. When $c \equiv E_g(1,r_1) \pmod n$,

$$\begin{aligned} c' &\equiv (g \times h^{r_1})^{m_1-m_0} \times g^{m_0} \pmod n \\ &\equiv g^{m_1-m_0} \times h^{r_1 \times (m_1-m_0)} \times g^{m_0} \pmod n \\ &\equiv g^{m_1} \times h^{r_1 \times (m_1-m_0)} \pmod n \\ &\equiv E_g(m_1, r_1 \times (m_1 - m_0)) \pmod n, \end{aligned}$$

which is a valid encryption of m_1. When $c \equiv E_g(0,r_0) \pmod n$,

$$\begin{aligned} c' &\equiv (h^{r_0})^{m_1-m_0} \times g^{m_0} \pmod n \\ &\equiv h^{r_0 \times (m_1-m_0)} \times g^{m_0} \pmod n \\ &\equiv E_g(m_0, r_0 \times (m_1 - m_0)) \pmod n, \end{aligned}$$

which is a valid encryption of m_0. So A calculates c', which turns out to be a valid encryption of m_0 or m_1. If c' is a valid encryption of m_1, then c is $E_g(1,r_1)$, and if c' is a valid encryption of m_0, then c is $E_g(0,r_0)$. Since A can distinguish between encryptions of m_1 and m_0, she can distinguish encryptions of 1 and 0. Therefore if the Okamoto-Uchiyama algorithm is not semantically secure then PSUB is not intractable, because solving PSUB is equivalent to distinguishing encryptions of 0 and 1 for this algorithm as explained before.

8.6 Example

Let $p = 7$, $q = 11$, $n = 7^2 \times 11 = 539$. The Sylow 7-subgroup H of $\mathbb{Z}_{49}^*$ is given by

$$H = \{1, 8, 15, 22, 29, 36, 43\}.$$

Every element in H, except 1, is eligible as g^{p-1}. Let $g = 8$, then

$$g^{p-1} \equiv 8^6 \equiv 43 \not\equiv 1 \pmod{49}$$

is a member of H. So $g = 8$ is eligible as a base. We get also $h \equiv g^n \equiv 8^{539} \equiv 491 \pmod{539}$ and $(L(43 \pmod{49}))^{-1} \equiv 6^{-1} \equiv 6 \pmod 7$. Let $m = 5$ with randomness $r = 13$. Then the encrypted message $c = E_g(m,r)$ is found as follows:

$$\begin{aligned} E_8(5,13) &\equiv 8^5 \times 491^{13} \pmod{539} \\ &\equiv 428 \times 442 \pmod{539} \\ &\equiv 526 \pmod{539}. \end{aligned}$$

Now the sender sends $c = 526$ to the receiver. Once the receiver gets c, the decryption works as follows:

$$\begin{aligned}
L(c^{p-1} \pmod{p^2}) \times (L(g^{p-1} \pmod{p^2}))^{-1} &\equiv \\
&\equiv L(526^6 \pmod{49}) \times 6 \pmod{7} \\
&\equiv L(15) \times 6 \pmod{7} \\
&\equiv \frac{15-1}{7} \times 6 \pmod{7} \\
&\equiv 2 \times 6 \pmod{7} \\
&\equiv 5 \pmod{7} \\
&\equiv m \pmod{7}.
\end{aligned}$$

To illustrate the homomorphic property, further let $m' = 1$ with randomness $r' = 15$. Then the encryption of m', which is c', is given by

$$\begin{aligned}
E_8(1,15) &\equiv 8^1 \times 491^{15} \pmod{539} \\
&\equiv 8 \times 197 \pmod{539} \\
&\equiv 498 \pmod{539}.
\end{aligned}$$

Now we calculate $E_g(m+m', r+r') = E_8(6,28)$.

$$\begin{aligned}
E_8(6,28) &\equiv 8^6 \times 491^{28} \pmod{539} \\
&\equiv 190 \times 295 \pmod{539} \\
&\equiv 533 \pmod{539}.
\end{aligned}$$

Also we calculate $E_g(m,r) \times E_g(m',r')$.

$$E_8(5,13) \times E_8(1,15) \equiv 526 \times 498 \equiv 533 \pmod{539},$$

and hence this verifies Eq. (8.4).

To exemplify the homomorphic multiplicative property by an integer scalar, use the message $m' = 1$ and the randomness $r' = 15$. We already know that $c' = E_8(1,15) = 498$. Consider $k' = 3$ as a scalar. We shall calculate the encryption of $k' \times m' \equiv 3 \pmod{539}$ with randomness $k' \times r' \equiv 45 \pmod{539}$. Then $E_g(k' \times m', k' \times r')$ is computed as

$$\begin{aligned}
E_8(3,45) &\equiv 8^3 \times 491^{45} \pmod{539} \\
&\equiv 512 \times 197 \pmod{539} \\
&\equiv 71 \pmod{539}.
\end{aligned}$$

Finally we compute $(c')^k \equiv 498^3 \equiv 71 \pmod{539}$, and this verifies Eq. (8.5).

To illustrate the other homomorphic property by an integer scalar, consider previously selected message $m = 5$ and randomness $r = 13$. Let $k = 2$ be a constant. Calculate

$$\begin{aligned} E_g(m+k,r) &= E_g(5+2,13) \\ &\equiv 8^7 \times 491^{13} \pmod{539} \\ &\equiv 442 \times 442 \pmod{539} \\ &\equiv 246 \pmod{539}. \end{aligned}$$

On the other hand, since $E_g(m,r) = 526$ by previous calculations,

$$g^k \times E_g(m,r) \equiv 8^2 \times 526 \equiv 64 \times 526 \equiv 246 \pmod{539},$$

which confirms Eq. (8.6).

To illustrate the re-randomization property, again take $m = 5$ and $r = 13$. Let $r'' = 4$ be a random number. Note that $r + r'' \equiv 13 + 4 \equiv 17 \pmod{539}$. From previous calculations we have $c \equiv E_8(5,13) \equiv 526 \pmod{539}$. Calculating

$$\begin{aligned} h^{r''} \times c &\equiv 491^4 \times 526 \pmod{539} \\ &\equiv 344 \times 526 \pmod{539} \\ &\equiv 379 \pmod{539}, \end{aligned}$$

and also

$$\begin{aligned} c'' &\equiv E_g(m, r+r'') \pmod{n} \\ &\equiv E_8(5,17) \pmod{539} \\ &\equiv 8^5 \times 491^{17} \pmod{539} \\ &\equiv 428 \times 50 \pmod{539} \\ &\equiv 379 \pmod{539}, \end{aligned}$$

we see that $D(491^4 \times E_8(5,13)) = D(E_8(5,13+4))$. Decrypting the re-randomized ciphertext $c'' = 379$, we recover the original message m:

$$\begin{aligned} D(379) &\equiv L(379^6 \pmod{49}) \times 6 \pmod{7} \\ &\equiv L(15) \times 6 \pmod{7} \\ &\equiv 2 \times 6 \pmod{7} \\ &\equiv 5 \pmod{7}. \end{aligned}$$

We see that $D(c) = D(c'')$, which confirms Eq. (8.7).

Chapter 9
Paillier Algorithm

The Paillier public-key algorithm was developed by Pascal Paillier [19] in 1999. The security of the Paillier algorithm relies on finding composite residuosity classes modulo n^2, where $n = p \times q$ and p, q are distinct primes, which is known as the Composite Residuosity Class Problem (Class$[n]$). (See Sect. 2.1.12.)

9.1 Key Generation

Key generation involves four steps.

(i) Choose distinct large k-bit primes p and q such that $\gcd(p, q-1) = 1$ and $\gcd(p-1, q) = 1$, and compute $n = p \times q$.
(ii) Compute $\lambda(n) = \text{lcm}(p-1, q-1)$.
(iii) Choose a semi-random base $g \in \mathbb{Z}^*_{n^2}$ such that $n | \text{ord}(g)$, where $\text{ord}(g)$ denotes the order of g in the cyclic group $\mathbb{Z}^*_{n^2}$.

The pair (n, g), where n is the modulus and g is the base of encryption, is released as a public key but the private decryption key $\lambda(n)$ is kept secret.

9.2 Encryption

After converting an alphanumeric plaintext message M into a purely numeric message m, where $0 \leq m < n$, the sender chooses a random $u \in \mathbb{Z}^*_{n^2}$ such that $0 < u < n$. In fact $u \in \mathbb{Z}^*_n$. Then the encrypted message c is computed as follows:

$$c \equiv g^m \times u^n \pmod{n^2}. \tag{9.1}$$

For each message to be encrypted, a different random u is chosen in order to ensure security.

Ç. K. Koç et al., *Partially Homomorphic Encryption*,
https://doi.org/10.1007/978-3-030-87629-6_9

Algorithm 30 Key generation algorithm for the Paillier cryptosystem

1: **Input:** Large odd distinct k-bit primes p,q such that $\gcd(p,q-1)=1$ and $\gcd(q,p-1)=1$.
2: **Output:** Public key (n,g) and secret key λ.
3: $n:=p\times q$
4: λ:=$(p-1)\times(q-1)$ ▷ Calculates secret key $\lambda(n)$.
5: **repeat**
6: g:=RANDINT$(1,n^2-1)$ ▷ Chooses encryption base.
7: g':=1 ▷ Initializes an element.
8: i:=0 ▷ Initializes an element.
9: **repeat**
10: $g'=g'\times g \pmod{n^2}$
11: $i=i+1$
12: **until** $g'=1$ ▷ Calculates ord(g) in $\mathbb{Z}^*_{n^2}$.
13: $r:=i \pmod n$ ▷ Checks whether ord(g) divides n.
14: **until** $r=0$ and GCD(g,n^2)=1
15: **return** (n,g) and λ

Algorithm 31 Encryption algorithm for the Paillier cryptosystem

1: **Input:** Message m, encryption key (n,g).
2: **Output:** Encrypted message c.
3: **repeat** ▷ Chooses randomness.
4: u:=RANDINT$(1,n-1)$
5: **until** GCD(u,n)=1
6: c:=LRPOW$(g,m,n^2)\times$ LRPOW$(u,n,n^2) \pmod{n^2}$
7: **return** c

Choice of Base g

Consider the encryption function

$$\begin{aligned} E_g(m,u) : \mathbb{Z}_n \times \mathbb{Z}_n^* &\to \mathbb{Z}^*_{n^2} \\ (m,u) &\mapsto g^m \times u^n. \end{aligned}$$

During the key generation process it is important to choose g such that $n|\operatorname{ord}(g)$, since in that case the encryption function is bijective. That is to say given any $w\in\mathbb{Z}^*_{n^2}$, with n fixed, the pair (m,u) is a unique pair such that $E_g(m,u)\equiv w$ $(mod\ n^2)$.

Lemma 14 *E_g is a bijection whenever $n|\operatorname{ord}(g)$.*

Proof. Since $|\mathbb{Z}_n\times\mathbb{Z}_n^*| = n\times\phi(n) = |\mathbb{Z}^*_{n^2}|$, it is sufficient to show that E_g is one-to-one. Assume $E_g(m_1,u_1)=E_g(m_2,u_2)$, where $m_1,m_2\in\mathbb{Z}_n$ and $u_1,u_2\in\mathbb{Z}_n^*$. Then

$$g^{m_1}\times(u_1)^n \equiv g^{m_2}\times(u_2)^n \pmod{n^2}.$$

Since $u_1,u_2\in\mathbb{Z}_n^*$ implies $u_1,u_2\in\mathbb{Z}^*_{n^2}$, their inverses exist modulo n^2. So

$$\begin{aligned} g^{m_1}\times(u_1)^n\times(g^{m_2}\times(u_2)^n)^{-1} &\equiv 1 \pmod{n^2} \\ g^{m_1-m_2}\times(u_1\times u_2^{-1})^n &\equiv 1 \pmod{n^2}. \end{aligned}$$

Raising the above equivalence to the power $\lambda(n)$, we get

$$g^{(m_1-m_2)\times\lambda(n)} \times (u_1 \times u_2^{-1})^{n\times\lambda(n)} \equiv 1 \pmod{n^2}.$$

Since $\lambda(n^2) = n \times \lambda(n)$ and $(u_1 \times u_2^{-1}) \in \mathbb{Z}_{n^2}^*$, by Carmichael's Theorem (Theorem 12) we have

$$(u_1 \times u_2^{-1})^{n\times\lambda(n)} \equiv 1 \pmod{n^2}.$$

So $g^{(m_1-m_2)\times\lambda(n)} \equiv 1 \pmod{n^2}$, which implies $\mathrm{ord}(g)|(\lambda(n) \times (m_1 - m_2))$. By assumption $n|\mathrm{ord}(g)$ so n divides $\lambda(n) \times (m_1 - m_2)$ as well. Since $\gcd(n, \lambda(n)) = 1$, we have $n|(m_1 - m_2)$, i.e. $m_1 \equiv m_2 \pmod{n}$. This means that $m_1 = m_2$ because $m_1, m_2 < n$. Turning back to the original equivalence

$$g^{(m_1-m_2)} \times (u_1 \times u_2^{-1})^n \equiv 1 \pmod{n^2},$$

and using $m_1 = m_2$, we get

$$\begin{aligned}(u_1 \times u_2^{-1})^n &\equiv 1 \pmod{n^2}\\ u_1^n &\equiv u_2^n \pmod{n^2}.\end{aligned}$$

The above equivalence is satisfied only if $u_1 \equiv u_2 \pmod{n}$. Consider $u_1 \equiv u_2 + \alpha \times n \pmod{n^2}$. Then

$$\begin{aligned}u_1^n &\equiv (u_2 + \alpha \times n)^n \pmod{n^2}\\ &\equiv u_2^n + \binom{n}{1} \times u_2^{n-1} \times (\alpha \times n) + \mathscr{O}(n^2) \pmod{n^2}\\ &\equiv u_2^n + u_2^{n-1} \times \alpha \times n^2 \pmod{n^2}\\ &\equiv u_2^n \pmod{n^2},\end{aligned}$$

where $\mathscr{O}(n^2)$ denotes the sum of the terms with higher powers of n and $0 \le \alpha < n$ is an integer. This gives all n solutions of the equivalence $u_1^n \equiv u_2^n \pmod{n^2}$. Since $u_1, u_2 < n$, we have $u_1 = u_2$. This completes the proof.

We continue with a definition from [19] which describes unique pairs (m, u) for a fixed n and a chosen g.

Definition 1 *Assume that $n|ord(g)$. For $w \in \mathbb{Z}_{n^2}^*$, we call the nth residuosity class of w with respect to g the unique integer $m \in \mathbb{Z}_n$ for which there exists $u \in \mathbb{Z}_n^*$ such that*

$$E_g(m, u) = w,$$

and we write $[\![w]\!]_g = m$ whenever $w = g^m \times u^n$ for some $u \in \mathbb{Z}_n^*$.

A more general version of Definition 1 is described in Theorem 20.

Well-Definedness of Logarithm Function

Let $\mathscr{S}_n$ be the subgroup of $\mathbb{Z}^*_{n^2}$ given by

$$\mathscr{S}_n = \{z < n^2 | z \equiv 1 \pmod{n}\}.$$

Since we have $(z-1)/n \in \mathbb{Z}$ for all $z \in \mathscr{S}_n$, $L(z) = (z-1)/n \pmod{n}$ is well-defined over $\mathscr{S}_n$.

9.3 Decryption

Once the receiver gets the encrypted message c, the original message m can be recovered using $\lambda(n)$ as follows:

$$m \equiv L(c^{\lambda(n)} \pmod{n^2}) \times (L(g^{\lambda(n)} \pmod{n^2}))^{-1} \pmod{n}. \tag{9.2}$$

The reader may see Lemma 16 to see why $L(g^{\lambda(n)} \pmod{n^2})$ has an inverse modulo n. The rest of this section is devoted to the proof of correctness of equivalence (9.2).

Algorithm 32 Algorithm for discrete logarithm of the Paillier cryptosystem

```
1: function DLFP(c,d,n)
2:     Input: Integers c, d and n, where d is the decryption key.
3:     Output: Logarithm of c.
4:     z:=LRPOW(c,d,n^2)
5:     u:=(z − 1)/n
6:     return u
```

Correctness of Decryption Algorithm

Before we start to discuss decryption, we need the following lemmas.

Lemma 15 *Consider* $(1+n) \in \mathbb{Z}^*_{n^2}$. *Then for any positive integer* ν

$$(1+n)^\nu \equiv 1 + \nu \times n \pmod{n^2}. \tag{9.3}$$

Proof.

$$
\begin{aligned}
(1+n)^{\nu} &\equiv 1+\binom{\nu}{1}\times n+\mathcal{O}(n^2) \pmod{n^2}\\
&\equiv 1+n\times\nu \pmod{n^2},
\end{aligned}
$$

where $\mathcal{O}(n^2)$ denotes the sum of the terms containing higher powers of n.

Note that

$$(1+n)^n \equiv 1 \pmod{n^2}$$

and for any degree ν less than n we have

$$(1+n)^{\nu} \equiv 1+\nu\times n \not\equiv 1 \pmod{n^2}.$$

As a consequence of Lemma 15, $\mathrm{ord}(1+n)=n$ in $\mathbb{Z}^*_{n^2}$ and $(1+n)^{-1}=(1+n)^{n-1}$. Note that among all elements of $\mathbb{Z}^*_{n^2}$ whose order is divisible by n, powers of $(1+n)$ are the easiest to perform algebraic operations on. We will prove the next lemma using this fact.

Lemma 16 *Let* $w\in\mathbb{Z}^*_{n^2}$*. Then*

$$L(w^{\lambda(n)} \pmod{n^2}) \equiv \lambda(n)\times[\![w]\!]_{(1+n)} \pmod{n}. \tag{9.4}$$

Proof. Since n divides $\mathrm{ord}(1+n)=n$, there exists a unique pair $(x,y)\in\mathbb{Z}_n\times\mathbb{Z}^*_n$ such that $w=(1+n)^x\times y^n \pmod{n^2}$ by Lemma 14. Hence $x=[\![w]\!]_{(1+n)}$. Raising the equivalence for w to the power $\lambda(n)$, by Carmichael's Theorem (Theorem 12) we get

$$
\begin{aligned}
w^{\lambda(n)} &\equiv (1+n)^{x\times\lambda(n)}\times y^{n\times\lambda(n)} \pmod{n^2}\\
&\equiv (1+n)^{x\times\lambda(n)} \pmod{n^2}\\
&\equiv 1+x\times n\times\lambda(n) \pmod{n^2}\\
&\equiv 1+\alpha\times n \pmod{n^2},
\end{aligned}
$$

where $\alpha\equiv x\times\lambda(n) \pmod{n}$ and $0\le\alpha<n$. Note that $1+\alpha\times n\in\mathcal{S}_n$. Hence

$$
\begin{aligned}
L(w^{\lambda(n)} \pmod{n^2}) &\equiv L(1+\alpha\times n) \pmod{n}\\
&\equiv \frac{\alpha\times n}{n} \pmod{n}\\
&\equiv \alpha \pmod{n}\\
&\equiv \lambda(n)\times[\![w]\!]_{(1+n)} \pmod{n},
\end{aligned}
$$

which completes the proof.

Lemma 17 *Let* $E_g(m,u)=c$*. Then*

$$m\equiv[\![c]\!]_{(1+n)}\times([\![g]\!]_{(1+n)})^{-1} \pmod{n}. \tag{9.5}$$

Proof. First note that $[\![g]\!]_{(1+n)}$ has an inverse, since $(1+n)$ and u are invertible. Suppose that $g \equiv (1+n)^{[\![g]\!]_{(1+n)}} \times (u')^n \pmod{n^2}$ for some $u' \in \mathbb{Z}_n^*$. Then $E_g(m,u) = c$ yields

$$\begin{aligned} c &\equiv g^m \times u^n \pmod{n^2} \\ &\equiv ((1+n)^{[\![g]\!]_{(1+n)}} \times (u')^n)^m \times u^n \pmod{n^2} \\ &\equiv (1+n)^{[\![g]\!]_{(1+n)} \times m} \times (u')^{m \times n} \times u^n \pmod{n^2} \\ &\equiv (1+n)^{[\![g]\!]_{(1+n)} \times m} \times ((u')^m \times u)^n \pmod{n^2}. \end{aligned}$$

Now let $u'' = (u')^m \times u$. Clearly $u'' \in \mathbb{Z}_n^*$ because $u', u \in \mathbb{Z}_n^*$. Recall that $[\![g]\!]_{(1+n)}$ and $[\![c]\!]_{(1+n)}$ are congruence classes modulo n. So

$$c = E_{(1+n)}(m \times [\![g]\!]_{(1+n)}, u''),$$

which means

$$[\![c]\!]_{(1+n)} \equiv m \times [\![g]\!]_{(1+n)} \pmod{n}.$$

Hence $m \equiv [\![c]\!]_{(1+n)} \times ([\![g]\!]_{(1+n)})^{-1} \pmod{n}$.

Using Lemma 16 and Lemma 17, we can prove that the decryption algorithm is correct.

Theorem 38 *Equivalence (9.2) is correct.*

Proof. Note that $\lambda(n) \times [\![g]\!]_{(1+n)}$ has an inverse, since both $\lambda(n)$ and $[\![g]\!]_{(1+n)}$ are invertible. Then $L(g^{\lambda(n)} \pmod{n^2})$ is invertible by Lemma 16.

$$\begin{aligned} & L(c^{\lambda(n)} \pmod{n^2}) \times (L(g^{\lambda(n)} \pmod{n^2}))^{-1} \\ \equiv\ & \lambda(n) \times [\![c]\!]_{(1+n)} \times (\lambda(n) \times [\![g]\!]_{(1+n)})^{-1} \pmod{n} \\ \equiv\ & [\![c]\!]_{(1+n)} \times ([\![g]\!]_{(1+n)})^{-1} \pmod{n} \\ \equiv\ & m \pmod{n}, \end{aligned}$$

where the last equivalence comes directly from Lemma 17.

Algorithm 33 Decryption algorithm for the Paillier cryptosystem

1: **Input:** Received message c, base of encryption g, decryption key d.
2: **Output:** Message m.
3: m:=DLFP(c,d,n)×INVERSE(DLFP(g,d,n),n) $\pmod{n}$
4: **return** m

9.4 Homomorphic Properties

Let $n = p \times q$ and $g \in \mathbb{Z}^*_{n^2}$ be chosen as the modulus and the base of encryption, respectively. Then the encryption function can be described as:

$$\begin{aligned} E_g : \mathbb{Z}_n \times \mathbb{Z}^*_n &\to \mathbb{Z}^*_{n^2} \\ (m,u) &\mapsto g^m \times u^n. \end{aligned}$$

Recall that the sender must change the random $u \in \mathbb{Z}^*_n$ accompanying m for every message to be encrypted. Denote the decryption function by D. Consider the messages m_1 and m_2 with their accompanying random u-values u_1 and u_2, respectively. Then the Paillier algorithm is considered to be additively homomorphic with the accompanying u-value of the message $m_1 + m_2$ being $u_1 \times u_2$, where $u_1 \times u_2 \in \mathbb{Z}^*_n$, since $u_1, u_2 \in \mathbb{Z}^*_n$.

$$\begin{aligned} E_g(m_1,u_1) \times E_g(m_2,u_2) &= (g^{m_1} \times (u_1)^n) \times (g^{m_2} \times (u_2)^n) \pmod{n^2} \\ &\equiv g^{m_1+m_2}(u_1 \times u_2)^n \pmod{n^2} \\ &= E_g(m_1+m_2, u_1 \times u_2), \end{aligned}$$

where the randomness in the encryption of $m_1 + m_2$ is $u_1 \times u_2$, which is neither uniformly distributed in $\mathbb{Z}^*_{n^2}$ nor independent of the randomness in $E_g(m_1,u_1)$ and $E_g(m_2,u_2)$. However this can be addressed by re-randomization, which is explained at the end of this section. This means that if we denote $E_g(m,u) = E(m)$, then for two messages m_1 and m_2 we have

$$D(E(m_1+m_2)) = D(E(m_1) \times E(m_2)). \tag{9.6}$$

On the other hand, no multiplicatively homomorphic operation $\boxtimes$ is known such that

$$D(E(m_1 \times m_2)) = D(E(m_1) \boxtimes E(m_2))$$

for the Paillier algorithm. So the Paillier algorithm is an additively homomorphic encryption scheme.

The Paillier algorithm also supports multiplication by a constant as follows:

$$\begin{aligned} (E_g(m,u))^k &= (g^m)^k \times (u^k)^n \pmod{n^2} \\ &\equiv g^{m\times k} \times (u')^n \pmod{n^2} \\ &= E_g(k \times m, u'), \end{aligned}$$

where $u' = u^k$ is a member of $\mathbb{Z}^*_n$, since $u \in \mathbb{Z}^*_n$. This means

$$D(E(k \times m)) = D((E(m))^k), \tag{9.7}$$

since $E_g(k \times m, u')$ is a valid encryption of $k \times m$. The same result can be calculated by Eq. (9.6).

Another homomorphic property supported by the Paillier algorithm is addition with a constant.

$$\begin{aligned} E_g(m+k,u) &= g^{m+k} \times (u)^n \pmod{n^2} \\ &\equiv (g^m \times g^k) \times u^n \pmod{n^2} \\ &\equiv (g^m \times u^n) \times g^k \pmod{n^2} \\ &\equiv c \times g^k \pmod{n^2}, \end{aligned}$$

which means

$$D(E(k+m)) = D(E(m) \times g^k). \tag{9.8}$$

Any ciphertext $c = E_g(m,u)$ can be re-randomized by calculating

$$\begin{aligned} E_g(m,u) \times (u')^n &\equiv c \times (u')^n \pmod{n^2} \\ &\equiv (g^m \times u^n) \times (u')^n \pmod{n^2} \\ &\equiv g^m \times (u \times u')^n \pmod{n^2} \\ &= E_g(m, u \times u'), \end{aligned}$$

where $u, u' \in \mathbb{Z}_n^*$. Note that $u \times u'$ is also a member of $\mathbb{Z}_n^*$, which makes $E_g(m, u \times u')$ a valid encryption of message m. Hence

$$D(E(m) \times (u')^n) = D(E(m)). \tag{9.9}$$

9.5 Security

The security of the Paillier algorithm relies on Class$[n]$.

Theorem 39 *If Class$[n]$ is hard in $\mathbb{Z}_{n^2}^*$ then the Paillier encryption function is not invertible by an adversary.*

Proof. Inverting Paillier's scheme is by definition Class$[n]$.

The semantic security of the Paillier algorithm relies on the Decisional Composite Residuosity Class Problem (D-Class$[n]$).

Theorem 40 *The Paillier algorithm is semantically secure if and only if D-Class$[n]$ is intractable.*

Proof. ($\Rightarrow$) Assume that D-Class$[n]$ is not intractable. So an adversary A can decide nth residues modulo n^2. Let m_0 and m_1 be two different known messages and c a random encryption of either m_0 or m_1. A computes

$$c_0 \equiv c \times g^{-m_0} \pmod{n^2}$$

and

$$c_1 \equiv c \times g^{-m_1} \pmod{n^2}.$$

If $c = E_g(m_0, u_0)$ for some $u_0 \in \mathbb{Z}_n^*$, then the first equivalence reduces to $(u_0)^n$ and c_0 is an nth residue modulo n^2 but c_1 is not with overwhelming probability. Similarly, if $c = E_g(m_1, u_1)$ for some $u_1 \in \mathbb{Z}_n^*$, then the second equivalence reduces to $(u_1)^n$ and c_1 is an nth residue modulo n^2 but c_0 is not with overwhelming probability. By deciding which one is an nth residue, A can decide decryption of c correctly with very high probability. Hence Paillier's scheme is not semantically secure if D-Class$[n]$ is not intractable.

($\Leftarrow$) Conversely, assume that Paillier's algorithm is not semantically secure. So the adversary can distinguish encryptions of m_0 and m_1. Given w and x, A should decide whether or not $[\![w]\!]_g = x$. Equivalently, A should decide whether there exists $u \in \mathbb{Z}_n^*$ such that $w \equiv g^x \times u^n \pmod{n^2}$ (or $w \times g^{-x} \equiv u^n \pmod{n^2}$) to solve D-Class$[n]$. The adversary should determine whether $w \times g^{-x}$ is an nth residue modulo n^2 by Proposition 2. A can compute $c' \equiv g^{m_0} \times w \times g^{-x} \pmod{n^2}$. Then c' is a valid encryption of m_0 if $w \times g^{-x}$ is an nth residue modulo n^2. A's ability to distinguish encryptions of m_0 helps the adversary to decide whether c' is a valid encryption of m_0. Hence A solves D-Class$[n]$.

Therefore the semantic security of Paillier's scheme is equivalent to solving D-Class$[n]$.

9.6 Example

Assume that the sender chooses $p = 7$, $q = 5$, and calculates $n = 35$, $n^2 = 1225$. Then the sender selects $g = 39$ from $\mathbb{Z}_{n^2}^*$ with order $\mathrm{ord}(g) = 210$, which is divisible by $n = 35$. Let the message to be sent be $m \equiv 18 \pmod{35}$. The sender chooses a random $u \in \mathbb{Z}_{n^2}^*$ such that $0 < u < 35 - n$. Suppose $u = 23$. Then the encrypted message is

$$\begin{aligned} c &= E_g(m, u) \\ &= E_{39}(18, 23) \\ &\equiv g^m \times u^n \pmod{n^2} \\ &\equiv 39^{18} \times 23^{35} \pmod{1225} \\ &\equiv 806 \times 557 \pmod{1225} \\ &\equiv 592 \pmod{1225}. \end{aligned}$$

The receiver needs $\lambda(n)$ and $(L(g^{\lambda(n)} \pmod{n^2}))^{-1} \pmod{n}$ to decrypt.

$$\lambda(n) = \lambda(7 \times 5) = \mathrm{lcm}(\phi(7), \phi(5)) = \mathrm{lcm}(6, 4) = 12.$$

Next, one has to calculate

$$\begin{aligned}(L(g^{\lambda(n)} \pmod{n^2}))^{-1} &\equiv (L(39^{12} \pmod{1225}))^{-1} \pmod{n}\\ &\equiv (L(946))^{-1} \pmod{35}\\ &\equiv \left(\frac{946-1}{35}\right)^{-1} \pmod{35}\\ &= 27^{-1} \pmod{35}\\ &= 13 \pmod{35}.\end{aligned}$$

After the receiver has the encrypted message c, $(L(g^{\lambda(n)} \pmod{n^2}))^{-1} \equiv 13 \pmod{n}$ and the private key $\lambda(n) = 12$, the message m is recovered as follows:

$$\begin{aligned}m &\equiv L(c^{\lambda(n)} \pmod{n^2}) \times (L(g^{\lambda(n)} \pmod{n^2}))^{-1} \pmod{n}\\ &\equiv L(592^{12} \pmod{1225}) \times 13 \pmod{35}\\ &\equiv L(1086) \times 13 \pmod{35}\\ &\equiv \frac{1086-1}{35} \times 13 \pmod{35}\\ &\equiv 31 \times 13 \pmod{35}\\ &\equiv 403 \pmod{35}\\ &\equiv 18 \pmod{35}.\end{aligned}$$

Consider $m_1 = 18$ with $u_1 = 23$. We have already calculated $c_1 = E_{39}(18,23) = 592$. To illustrate the additive homomorphic property of the algorithm, further let $m_2 = 13$ and $u_2 = 19$. Then

$$\begin{aligned}c_2 &= E_{39}(13,19)\\ &\equiv 39^{13} \times 19^{35} \pmod{1225}\\ &\equiv 144 \times 374 \pmod{1225}\\ &\equiv 1181 \pmod{1225}.\end{aligned}$$

Let $m' = m_1 + m_2 = 18 + 13 = 31 \pmod{35}$ with $u' = u_1 \times u_2 = 23 \times 19 = 17 \pmod{35}$. Then

$$\begin{aligned}c' &= E_{39}(31,17)\\ &\equiv 39^{31} \times 17^{35} \pmod{1225}\\ &\equiv 914 \times 68 \pmod{1225}\\ &\equiv 902 \pmod{1225}.\end{aligned}$$

Note that

$$\begin{aligned} c' &\equiv c_1 \times c_2 \pmod{n^2} \\ &\equiv E_{39}(18,23) \times E_{39}(13,19) \pmod{1225} \\ &\equiv 592 \times 1181 \pmod{1225} \\ &\equiv 902 \pmod{1225}, \end{aligned}$$

which confirms $E_g(m_1 + m_2, u_1 \times u_2) = E_g(m_1, u_1) \times E_g(m_2, u_2)$.

To illustrate that the algorithm supports the multiplication by a constant, consider again the message $m = 18$ with $u = 23$ and a constant $k = 3$. We already have $c = E_{39}(18,23) = 592$ as the encryption of m. Let D be the decryption function. Then

$$\begin{aligned} D(c^k) &= D(592^3 \pmod{1225}) \\ &= D(113) \\ &\equiv L(113^{12} \pmod{1225}) \times 13 \pmod{35} \\ &\equiv \frac{806-1}{35} \times 13 \pmod{35} \\ &\equiv 19 \pmod{35}. \end{aligned}$$

On the other hand $k \times m = 3 \times 18 \equiv 19 \pmod{35}$. Hence c^k is a valid encryption of $k \times m$, which confirms Eq. (9.7).

Next, to illustrate that the algorithm supports the addition by a constant, consider the previously used m, u and k. Then

$$\begin{aligned} E_{39}(18+3,23) &\equiv 39^{21} \times 23^{35} \pmod{1225} \\ &\equiv 589 \times 557 \pmod{1225} \\ &\equiv 998 \pmod{1225}. \end{aligned}$$

We also get $E_{39}(18,23) \times 39^3 \equiv 998 \pmod{1225}$. This confirms Eq. (9.8).

Consider $u''' = 3$ for re-randomization of $c = E_g(m,u)$. Note that $u \times u''' \equiv 23 \times 3 \equiv 34 \pmod{35}$. Calculating

$$\begin{aligned} 3^{35} \times E_{39}(18,23) &\equiv 3^{35} \times 39^{18} \times 23^{35} \pmod{1225} \\ &\equiv 3^{35} \times 592 \pmod{1225} \\ &\equiv 607 \times 592 \pmod{1225} \\ &\equiv 419 \pmod{1225}, \end{aligned}$$

and also

$$\begin{aligned} E_{39}(18,34) &\equiv 39^{18} \times 34^{35} \pmod{1225} \\ &\equiv 806 \times 1224 \pmod{1225} \\ &\equiv 419 \pmod{1225}, \end{aligned}$$

we see that $3^{35} \times E_{39}(18,23) \equiv E_{39}(18, 23 \times 3) \pmod{1225}$, confirming Eq. (9.9).

Chapter 10
Damgård-Jurik Algorithm

The Damgård-Jurik algorithm was developed as a generalization of the Paillier algorithm by Ivan Damgård and Mads Jurik in 2001 [5]. This algorithm extends Paillier's scheme to $\mathbb{Z}^*_{n^{s+1}}$ in a very natural manner. As with Paillier's algorithm, the security of the Damgård-Jurik algorithm is based on the Composite Residuosity Class Problem (Class$[n]$). (See Sect. 2.1.12.)

10.1 Forming the Plaintext Space

Let $n = p \times q$ and $\lambda(n) = \text{lcm}(p-1, q-1)$, where p and q are prime numbers and λ is Carmichael's lambda function. Here $\gcd(n, \lambda(n)) = 1$ is satisfied except with negligible probability. Such an n is called **admissible** by Jurik and we refer to it in this way throughout this chapter. As in Paillier's scheme, the element $(1+n)$ has a key role in encryption and decryption.

Lemma 18 *For any admissible n and $s < p, q$, the element $(1+n)$ has order n^s in $\mathbb{Z}^*_{n^{s+1}}$.*

Proof. Consider the ith power of $(1+n)$ modulo n^{s+1}.

$$\begin{aligned}(1+n)^i &\equiv 1 + \binom{i}{1} \times n + \binom{i}{2} \times n^2 + \cdots + \binom{i}{i-1} \times n^{i-1} + n^i \pmod{n^{s+1}} \\ &\equiv 1 + n \times \left[\binom{i}{1} + \binom{i}{2} \times n + \cdots + \binom{i}{i-1} \times n^{i-2} + n^{i-1}\right] \pmod{n^{s+1}},\end{aligned}$$

which implies $\sum_{j=0}^{i} \binom{i}{j} \times n^j \equiv 1 \pmod{n^{s+1}}$ if and only if $\sum_{j=1}^{i} \binom{i}{j} \times n^{j-1} \equiv 0 \pmod{n^s}$. Let $a = \text{ord}(1+n)$ in $\mathbb{Z}^*_{n^{s+1}}$. If $i = n^s$, then

$$(1+n)^{n^s} \equiv 1 + n \times \left[\binom{n^s}{1} + \binom{n^s}{2} \times n + \cdots + \binom{n^s}{n^s-1} \times n^{n^s-2} + n^{n^s-1}\right] \pmod{n^{s+1}}.$$

Ç. K. Koç et al., *Partially Homomorphic Encryption*,
https://doi.org/10.1007/978-3-030-87629-6_10

Note that $\sum_{i=1}^{n^s} \binom{n^s}{i} \times n^{i-1} \equiv 0 \pmod{n^s}$ is true, because each term $\binom{n^s}{i} \times n^{i-1}$ is divisible by n^s for $1 \le i \le n^s$. This leads to $(1+n)^{n^s} \equiv 1 \pmod{n^{s+1}}$. It turns out that $a|n^s$, where $n^s = p^s \times q^s$. Set $\mathrm{ord}(1+n) = a = p^\alpha \times q^\beta$, where $\alpha, \beta \le s$. Consider

$$\begin{aligned}(1+n)^a - 1 &= \sum_{j=1}^{a} \binom{a}{j} \times n^{j-1} \\ &= \binom{a}{1} + \binom{a}{2} \times n + \cdots + \binom{a}{a-1} \times n^{a-2} + n^{a-1}.\end{aligned}$$

For $j > s$, $n^j > n^s$ is true, which means $n^s | \binom{a}{j} \times n^{j-1}$ and hence $a | \binom{a}{j} \times n^{j-1}$. For $j \le s$, $j!$ cannot have p or q as prime factors, since $j \le s < p, q$, which means $\binom{a}{j}$ is divisible by a. So for all cases $\binom{a}{j} \times n^{j-1}$ is divisible by a. Assume for a contradiction that $a = p^\alpha \times q^\beta < n^s$. Without loss of generality, as $\alpha < s$, we know that n^s divides $\sum_{j=1}^{a} \binom{a}{j} \times n^{j-1}$. Since p divides $n^s/a = p^{s-\alpha} \times q^{s-\beta}$, p must also divide $\sum_{j=1}^{a} \binom{a}{j} \times n^{j-1}/a$. However,

$$\sum_{j=1}^{a} \frac{\binom{a}{j} \times n^{j-1}}{a} = 1 + \frac{\binom{a}{2} \times n}{a} + \cdots + \frac{\binom{a}{a-1} \times n^{a-2}}{a} + \frac{n^{a-1}}{a},$$

and all the terms except the first one are divisible by p. Hence $p|1$, which leads to a contradiction.

The following proposition is a consequence of Carmichael's Theorem (Thm. 12).

Proposition 5 *For any $w \in \mathbb{Z}_{n^{s+1}}^*$*

$$w^{n^s \times \lambda(n)} \equiv 1 \pmod{n^{s+1}}.$$

Proof.

$$\begin{aligned}\lambda(n^{s+1}) &= \lambda(p^{s+1} \times q^{s+1}) \\ &= \mathrm{lcm}(\phi(p^{s+1}), \phi(q^{s+1})) \\ &= \mathrm{lcm}(p^s \times (p-1), q^s \times (q-1)) \\ &= p^s \times q^s \times \mathrm{lcm}(p-1, q-1) \\ &= n^s \times \lambda(n),\end{aligned}$$

which leads to $w^{n^s \times \lambda(n)} \equiv 1 \pmod{n^{s+1}}$ by Carmichael's Theorem (Theorem 12).

The following lemma enables us to express $\mathbb{Z}_{n^{s+1}}^*$ as a direct product. Also it characterizes the encryption function.

Lemma 19 *For any admissible n and $s < p, q$, the map*

$$\begin{aligned}E_s : \mathbb{Z}_{n^s} \times \mathbb{Z}_n^* &\to \mathbb{Z}_{n^{s+1}}^* \\ (m, u) &\mapsto (1+n)^m \times u^{n^s}\end{aligned}$$

is an isomorphism, where $E_s(m_1+m_2,u_1\times u_2)=E_s(m_1,u_1)\times E_s(m_2,u_2)$.

Proof. We begin with proving E_s is a homomorphism. Here $\mathbb{Z}_{n^s}$ is an additive group, and $\mathbb{Z}_n^*$ and $\mathbb{Z}_{n^{s+1}}^*$ are multiplicative ones. Let $m_1,m_2\in\mathbb{Z}_{n^s}$ and $u_1,u_2\in\mathbb{Z}_n^*$. Then

$$\begin{aligned}E_s(m_1+m_2,u_1\times u_2) &\equiv (1+n)^{m_1+m_2}\times(u_1\times u_2)^{n^s} \pmod{n^{s+1}}\\ &\equiv (1+n)^{m_1}\times u_1^{n^s}\times(1+n)^{m_2}\times u_2^{n^s} \pmod{n^{s+1}}\\ &\equiv E_s(m_1,u_1)\times E_s(m_2,u_2).\end{aligned}$$

Hence E_s is a homomorphism from $\mathbb{Z}_{n^s}\times\mathbb{Z}_n^*$ to $\mathbb{Z}_{n^{s+1}}^*$. Next we prove that E_s is bijective. It is sufficient to show that E_s is injective, since the order of $\mathbb{Z}_{n^s}\times\mathbb{Z}_n^*$ is the same as the order of $\mathbb{Z}_{n^{s+1}}^*$, which is equal to $n^s\times\lambda(n)$. Assume that $E_s(m_1,u_1)=E_s(m_2,u_2)$, i.e.

$$(1+n)^{m_1}\times u_1^{n^s}\equiv(1+n)^{m_2}\times u_2^{n^s} \pmod{n^{s+1}}.$$

Since $(1+n),u_1,u_2$ are all units, they have inverses in $\mathbb{Z}_{n^{s+1}}^*$. So we can write

$$(1+n)^{m_1-m_2}\times(u_1\times u_2^{-1})^{n^s}\equiv 1 \pmod{n^{s+1}}. \tag{10.1}$$

Raising the above equivalence to the power $\lambda(n)$, we get

$$(1+n)^{(m_1-m_2)\times\lambda(n)}\times(u_1\times u_2^{-1})^{n^s\times\lambda(n)}\equiv 1 \pmod{n^{s+1}},$$

and using Proposition 5 we have

$$(1+n)^{(m_1-m_2)\times\lambda(n)}\equiv 1 \pmod{n^{s+1}}.$$

Since $\mathrm{ord}(1+n)=n^s$ and $\lambda(n)=\mathrm{lcm}(p-1,q-1)$, $\gcd(\mathrm{ord}(1+n),\lambda(n))=1$ due to n being admissible by assumption. By the above equivalence we have $n^s|(m_1-m_2)\times\lambda(n)$, which means $n^s|(m_1-m_2)$. Here m_1 and m_2 are congruence classes modulo n^s. Hence $m_1=m_2$.

Turning back to the equivalence (10.1) and putting $m_1=m_2$, we get $(u_1\times u_2^{-1})^{n^s}\equiv 1 \pmod{n^{s+1}}$. But this happens only if $u_1\equiv u_2 \pmod{n}$. Let $u_1\equiv u_2+\alpha n \pmod{n^{s+1}}$ for some integer $0\le\alpha<n^s$, then

$$\begin{aligned}u_1^{n^s} &\equiv (u_2+\alpha\times n)^{n^s} \pmod{n^{s+1}}\\ &\equiv u_2^{n^s}+\binom{n^s}{1}\times u_2^{n^s-1}\times(\alpha\times n)+\cdots+\binom{n^s}{n^s-1}\times u_2\times(\alpha\times n)^{n^s-1}\\ &\quad+(\alpha\times n)^{n^s} \pmod{n^{s+1}}.\end{aligned}$$

Since $n^{s+1}|n^{n^s}$, we have $n^{n^s}\equiv 0 \pmod{n^{s+1}}$. For $1\le i\le n^s-1$, n^s appears as a factor in $\binom{n^s}{i}$ and n appears as a factor in $(\alpha\times n)^i$. Hence all terms $\binom{n^s}{i}\times u_2^{n^s-i}\times(\alpha\times n)^i$ are divisible by n^{s+1}. This gives all n^s solutions of the equivalence $u_1^{n^s}\equiv u_2^{n^s} \pmod{n^{s+1}}$. Since $u_1,u_2<n$, we have $u_1=u_2$. This completes the proof. Therefore

$(m_1, u_1) = (m_2, u_2)$. Finally we conclude that E_s is injective and therefore it is an isomorphism.

Given the factorization of n, E_s can be inverted. This means that for any element $g \in \mathbb{Z}^*_{n^{s+1}}$, there exists a unique pair $(m, u) \in \mathbb{Z}_{n^s} \times \mathbb{Z}^*_n$ such that $g = (1+n)^m \times u^{n^s}$. By letting $u = 1$, $\pi := E_s(m, 1)$ defines an isomorphism between $\mathbb{Z}_{n^s}$ and $G =< (1+n) >$, the subgroup of $\mathbb{Z}^*_{n^{s+1}}$ generated by $(1+n)$ with $\text{ord}(1+n) = n^s$. Hence the factor group $\mathbb{Z}^*_{n^{s+1}}/\mathbb{Z}^*_n \cong \mathbb{Z}_{n^s}$ is isomorphic to G. Denote $\mathbb{Z}^*_{n^{s+1}}/\mathbb{Z}^*_n = \overline{G}$, which consists of cosets of $\mathbb{Z}^*_n$ in $\mathbb{Z}^*_{n^{s+1}}$, namely

$$\begin{aligned}\overline{G} &= \{\mathbb{Z}^*_n, (1+n)\mathbb{Z}^*_n, (1+n)^2\mathbb{Z}^*_n, \cdots, (1+n)^{n^s-1}\mathbb{Z}^*_n\} \\ &= \{(1+n)^j\mathbb{Z}^*_n : j \in \mathbb{Z}_{n^s}\}.\end{aligned}$$

Note that $(1+n)\mathbb{Z}^*_n$ is a generator of $\overline{G}$ and there is a natural enumeration for elements of $\overline{G}$ coming from the exponents of $(1+n)$ describing the cosets. Exponents of these coset representatives form the plaintext space of this algorithm.

10.2 Key Generation

Key generation involves three steps.

(i) Choose an admissible $n = p \times q$ with distinct large primes p, q, and compute $\lambda(n) = \text{lcm}(p-1, q-1)$.
(ii) Choose a base $g \equiv (1+n)^j \times x \pmod{n^{s+1}}$ from $\mathbb{Z}^*_{n^{s+1}}$, for some j such that $\gcd(j, n) = 1$ and $x \in \mathbb{Z}^*_n$.
(iii) Choose d such that $d \pmod{n} \in \mathbb{Z}^*_n$, and $d \equiv 0 \pmod{\lambda(n)}$.

The pair (n, g) is released as the public encryption key and the private decryption key d is kept secret.

10.3 Encryption

After converting an alphanumeric plaintext message M into a purely numeric message m, where $m \in \mathbb{Z}_{n^s}$, the sender chooses a random $u \in \mathbb{Z}^*_{n^{s+1}}$ such that $0 < u < n$. In fact $u \in \mathbb{Z}^*_n$. Then the encrypted message c is computed as follows:

$$c \equiv g^m \times u^{n^s} \pmod{n^{s+1}}.$$

For each message to be encrypted, a different randomness u is chosen. Here we may denote the encryption function with base g by $E_{s,g}$. Therefore the encrypted message of the original message m with base g and randomness u is given by

$$c = E_{s,g}(m, u). \tag{10.2}$$

Algorithm 34 Key generation algorithm for the Damgård-Jurik cryptosystem

1: **Input:** Large odd distinct k-bit primes p,q.
$\triangleright$ $\gcd(p,q-1)=1$ and $\gcd(q,p-1)=1$ with a very high probability.
2: **Output:** Public key (n,g) and secret key d.
3: $n:=p\times q$
4: $\lambda:=(p-1)\times(q-1)$ $\triangleright$ Calculates $\lambda(n)$.
5: **repeat**
6: j:=RANDINT$(1,n^s-1)$
7: x:=RANDINT$(1,n-1)$
8: **until** GCD(j,n)= 1 and GCD(x,n)= 1
9: g:=LRPOW$(1+n,j,n^{s+1})\times x \pmod{n^{s+1}}$ $\triangleright$ Chooses the encryption base g.
10: **repeat**
11: d:=RANDINT$(1,n-1)$ $\triangleright$ Chooses the decryption key d.
12: d':=$d \pmod{\lambda}$
13: **until** GCD(d,n)= 1 and $d'=0$
14: **return** (n,g) and d

Algorithm 35 Encryption algorithm for the Damgård-Jurik cryptosystem

1: **Input:** Message m, encryption key (n,g), integer s.
2: **Output:** Encrypted message c.
3: **repeat** $\triangleright$ Chooses randomness.
4: u:=RANDINT$(1,n-1)$
5: **until** GCD(u,n)= 1
6: c:=LRPOW$(g,m,n^{s+1})\times$ LRPOW$(u,n^s,n^{s+1}) \pmod{n^{s+1}}$
7: **return** c

10.4 Decryption

Once the receiver gets the encrypted message c, the original message m is recovered as follows:

$$m \equiv L(c^d \pmod{n^{s+1}}) \times (L(g^d \pmod{n^{s+1}}))^{-1} \pmod{n^s}. \tag{10.3}$$

Well-Definedness of L

We define the logarithmic function L as:

$$L(z) = \frac{z-1}{n} \pmod{n^s}.$$

Let $z \in G$, so $z \equiv (1+n)^j \pmod{n^{s+1}}$ for some $j \in \mathbb{Z}_{n^s}$.

$$
\begin{aligned}
z - 1 &\equiv (1+n)^j - 1 \pmod{n^{s+1}} \\
&\equiv \left[1 + \binom{j}{1} \times n + \mathcal{O}(n^2)\right] - 1 \pmod{n^{s+1}} \\
&\equiv \binom{j}{1} \times n + \mathcal{O}(n^2) \pmod{n^{s+1}},
\end{aligned}
$$

where $\mathcal{O}(n^2)$ denotes the sum of the terms containing higher powers of n. Hence $z-1$ is divisible by n, which means $(z-1)/n$ is an integer. Therefore L is well-defined on G.

Algorithm 36 Algorithm for function L

```
1: function LDJ(c,n,j)
2:     Input: Integers c, j and n.
3:     Output: L(c).
4:     z:=c mod n^{j+1}
5:     u:=(z − 1)/n
6:     return u
```

Correctness of Decryption Algorithm

Theorem 41 *Equivalence (10.3) is correct.*

Proof. Given a ciphertext $c = E_{s,g}(m,u)$, the receiver first computes $c^d \pmod{n^{s+1}}$, where d is the private decryption key, as follows:

$$
\begin{aligned}
c^d &\equiv (g^m \times u^{n^s})^d \pmod{n^{s+1}} \\
&\equiv ((1+n)^j \times x)^{m \times d} \times u^{n^s \times d} \pmod{n^{s+1}} \\
&\equiv (1+n)^{j \times m \times d \pmod{n^s}} \times (x^m \times u^{n^s})^{d \pmod{\lambda(n)}} \pmod{n^{s+1}} \\
&\equiv (1+n)^{j \times m \times d \pmod{n^s}} \pmod{n^{s+1}},
\end{aligned}
$$

where the last two congruences come from the fact that $\mathrm{ord}(1+n) = n^s$ and that $x, u \in \mathbb{Z}_n^*$, $x^m \times u^{n^s} \in \mathbb{Z}_n^*$, $|\mathbb{Z}_n^*| = \lambda(n)$ and $d \equiv 0 \pmod{\lambda(n)}$. Then the receiver computes the exponent $j \times m \times d \pmod{n^s}$ by using the iterative method described in the next part. Also applying the same method on g produces the exponent $j \times d \pmod{n^s}$. Namely,

$$
\begin{aligned}
g^d &\equiv ((1+n)^j \times x)^d \pmod{n^{s+1}} \\
&\equiv (1+n)^{j \times d} \times x^d \pmod{n^{s+1}} \\
&\equiv (1+n)^{j \times d \pmod{n^s}} \times x^{d \pmod{\lambda(n)}} \pmod{n^{s+1}} \\
&\equiv (1+n)^{j \times d \pmod{n^s}} \pmod{n^{s+1}}.
\end{aligned}
$$

Recall that $\gcd(j,n) = 1$ and d is equivalent to a unit modulo n. Hence j and d have inverses modulo n^s. Therefore

$$(j \times m \times d) \times (j \times d)^{-1} \equiv m \pmod{n^s}.$$

Equivalence 10.3 follows.

Finding the Exponent of $(1+n)^i \pmod{n^{s+1}}$

Recall that raising c and g to the power d reduces the decryption process to finding powers of $(1+n)$ modulo n^s. Clearly,

$$\begin{aligned} L((1+n)^i \pmod{n^{s+1}}) &\equiv \frac{1+\binom{i}{1}\times n+\cdots+\binom{i}{s}\times n^s - 1}{n} \pmod{n^s} \\ &\equiv i+\binom{i}{2}\times n+\cdots+\binom{i}{s}\times n^{s-1} \pmod{n^s}. \end{aligned}$$

The algorithm for computing i works in an iterative manner. First we extract $i_1 \equiv i \pmod{n}$, then $i_2 \equiv i \pmod{n^2}$ and in general $i_j \equiv i \pmod{n^j}$. We extract $i_j \equiv i \pmod{n^j}$ as follows:

$$L((1+n)^i \pmod{n^{j+1}}) \equiv i_j + \binom{i_j}{2}\times n+\cdots+\binom{i_j}{j}\times n^{j-1} \pmod{n^j}. \quad (10.4)$$

It is easy to find i_1 as follows:

$$i_1 \equiv L((1+n) \pmod{n^2}) \equiv i \pmod{n}.$$

Now we need a way to connect the initial term i_1 to the other terms. In general we need a relation between i_{j-1} and i_j. Note that $i_j = i_{j-1} + k \times n^{j-1}$ for some $0 \leq k < n$, since $i \equiv i_{j-1} \equiv i_j \pmod{n^{j-1}}$. Then the right-hand side of equivalence (10.4) becomes

$$\begin{aligned} &\binom{i_{j-1}+k\times n^{j-1}}{1} + \binom{i_{j-1}+k\times n^{j-1}}{2} \times n + \cdots \\ &\cdots + \binom{i_{j-1}+k\times n^{j-1}}{j} \times n^{j-1} \pmod{n^j}. \end{aligned}$$

Consider the expansion of the second term.

$$\binom{i_{j-1}+k\times n^{j-1}}{2} \times n = \frac{(i_{j-1}+k\times n^{j-1})\times(i_{j-1}+k\times n^{j-1}-1)\times n}{2}.$$

Therefore we have

$$\binom{i_j}{2} \times n \equiv \binom{i_{j-1} + k \times n^{j-1}}{2} \times n \equiv \binom{i_{j-1}}{2} \times n \pmod{n^j}.$$

In general, for $j > t > 0$, this works as follows:

$$\binom{i_j}{t+1} \times n^t \equiv \binom{i_{j-1}}{t+1} \times n^t \pmod{n^j}.$$

Assuming that we already have i_{j-1}, we can isolate i_j by performing the following subtraction:

$$i_j \equiv L((1+n)^i \pmod{n^{j+1}}) - (\binom{i_j}{2} \times n + \cdots + \binom{i_j}{j} \times n^{j-1}) \pmod{n^j} \quad (10.5)$$

The reader may find an application of the algorithm in the example at the end of this chapter.

Algorithm 37 Algorithm for discrete logarithm in base $(1+n)$

```
 1: function DLFDJ(c,n,s)
 2:     Input: Integers c, s and n.
 3:     Output: Logarithm of c in base (1+n).
 4:     i:=0                                          ▷ Initializing a variable.
 5:     for j:=1 to s do                              ▷ i = i_{j-1}
 6:         h1:=LDJ(c,n,j)
 7:         h2:=i
 8:         for k:=2 to j do                          ▷ h2 := i × (i−1)···(i−k+2)
 9:             i := i − 1
10:             h2 := h2 (mod n^j)
11:             h1 := h1 − (h2 × n^{k−1})/k! (mod n^j)  ▷ h1 := h1 − binom(i,k) × n^k
12:         i := h1
13:     return i
```

Algorithm 38 Decryption algorithm for the Damgård-Jurik cryptosystem

```
1: Input: Received message c, s, decryption key d, n, g encryption base.
2: Output: Original message m.
3: a:=LRPOW(c,d,n^{s+1})
4: j:=DLFDJ(a,n,s)
5: b:=LRPOW(g,d,n^{s+1})
6: j':=DLFDJ(b,n,s)
7: j'':=INVERSE(j',n^s)
8: m:=j × j'' (mod n^s)
9: return m
```

10.5 Homomorphic Properties

As a generalization of the Paillier algorithm, Damgård-Jurik inherits all the homomorphic properties of the Paillier algorithm.

Let $n = p \times q$ and $g \in \mathbb{Z}^*_{n^{s+1}}$ with $g \equiv (1+n)^j \times x \pmod{n^{s+1}}$ be chosen as the modulus and the base of encryption, respectively. Then the encryption function can be described as:

$$\begin{aligned} E_{s,g} : \mathbb{Z}_{n^s} \times \mathbb{Z}^*_n &\to \mathbb{Z}^*_{n^{s+1}} \\ (m,u) &\mapsto g^m \times u^{n^s}. \end{aligned}$$

Recall that the sender must change the random $u \in \mathbb{Z}^*_n$ accompanying m for every message to be encrypted. Let D be the decryption function. Consider the messages m_1 and m_2 with their accompanying random values u_1 and u_2, respectively. Then the Damgård-Jurik algorithm is considered to be additively homomorphic with the accompanying random value of the message $m_1 + m_2$ being $u_1 \times u_2$, where $u_1 \times u_2 \in \mathbb{Z}^*_n$, since $u_1, u_2 \in \mathbb{Z}^*_n$.

$$\begin{aligned} E_{s,g}(m_1,u_1) \times E_{s,g}(m_2,u_2) &= (g^{m_1} \times (u_1)^{n^s}) \times (g^{m_2} \times (u_2)^{n^s}) \pmod{n^{s+1}} \\ &\equiv g^{m_1+m_2} \times (u_1 \times u_2)^{n^s} \pmod{n^{s+1}} \\ &= E_{s,g}(m_1 + m_2, u_1 \times u_2), \end{aligned}$$

where the randomness in the encryption of $m_1 + m_2$ is $u_1 \times u_2$, which is neither uniformly distributed in $\mathbb{Z}^*_{n^{s+1}}$ nor independent of the randomness in $E_{s,g}(m_1,u_1)$ and $E_{s,g}(m_2,u_2)$. However this can be addressed by re-randomization, which is explained at the end of this section. This means that if we denote $E_{s,g}(m,u) = E(m)$, then for two messages m_1 and m_2 we have

$$D(E(m_1+m_2)) = D(E(m_1) \times E(m_2)). \tag{10.6}$$

On the other hand, no multiplicatively homomorphic operation $\boxtimes$ is known such that

$$D(E(m_1 \times m_2)) = D(E(m_1) \boxtimes E(m_2))$$

for this algorithm. So Damgård-Jurik is a partially homomorphic encryption scheme. This algorithm also supports multiplication by a constant as follows:

$$\begin{aligned} (E_{s,g}(m,u'))^k &= (g^m)^k \times ((u')^k)^{n^s} \pmod{n^{s+1}} \\ &\equiv g^{m \times k} \times (u)^{n^s} \pmod{n^{s+1}} \\ &= E_{s,g}(k \times m, u), \end{aligned}$$

where $u = (u')^k$ is a member of $\mathbb{Z}^*_n$, since $u' \in \mathbb{Z}^*_n$. This means

$$D(E(k \times m)) = D((E(m))^k), \tag{10.7}$$

since $E_{s,g}(m,u')$ is a valid encryption of message m. Indeed this property follows from equivalence (10.6).

Another homomorphic property supported by the Damgård-Jurik algorithm is addition with a constant.

$$\begin{aligned}
c \times g^k &= (g^m \times u^{n^s}) \times g^k \pmod{n^{s+1}} \\
&= g^{m+k} \times (u)^{n^s} \pmod{n^{s+1}} \\
&= (g^m \times g^k) \times u^{n^s} \pmod{n^{s+1}} \\
&= E_{s,g}(m+k,u) \pmod{n^{s+1}},
\end{aligned}$$

which means

$$D(E(k+m)) = D(E(m) \times g^k). \tag{10.8}$$

Re-randomization of messages is also possible. If u' is a random element of $\mathbb{Z}_n^*$, then

$$c \times (u')^{n^s} \equiv g^m \times u^{n^s} \times (u')^{n^s} \equiv g^m \times (u \times u')^{n^s} \pmod{n^{s+1}},$$

which is a valid encryption of m, since $u \times u'$ is a member of $\mathbb{Z}_n^*$. Hence

$$D(c) = D(c \times (u')^{n^s}). \tag{10.9}$$

10.6 Security

The security of the Damgård-Jurik algorithm relies on Class$[n]$.

Theorem 42 *If Class$[n]$ is hard in $\mathbb{Z}_{n^{s+1}}^*$ then the Damgård-Jurik encryption function is not invertible by an adversary.*

Proof. Inverting the Damgård-Jurik scheme is by definition solving Class$[n]$.

The semantic security of the Damgård-Jurik algorithm relies on the Decisional Composite Residuosity Class Problem (D-Class$[n]$), which was introduced by Paillier in [19].

Theorem 43 *The Damgård-Jurik algorithm is semantically secure if and only if D-Class$[n]$ is intractable.*

Proof. ($\Rightarrow$) Assume that D-Class$[n]$ is not intractable and let c be a ciphertext modulo n^{s+1}. Then an adversary A who can solve the decisional composite residuosity problem can reduce c modulo n^2, m modulo n, and get a ciphertext from the Paillier cryptosystem with non-negligible probability. Thus, A can break the semantic security of the Damgård-Jurik cryptosystem.

($\Leftarrow$) Conversely, we will prove that the security of the Damgård-Jurik cryptosystem follows from the intractibility of D-Class$[n]$ for any s by induction on s. Clearly, for $s = 1$, the security comes from the Paillier algorithm's security. Suppose that

$s > 1$ and for any $t < s$ the Damgård-Jurik cryptosystem is semantically secure. Assume for a contradiction that the Damgård-Jurik cryptosystem is not secure for $s = t+1$. Then given two messages $m_0, m_1 \in \mathbb{Z}_{n^t}$, A can distinguish encryptions of m_0 and m_1. Let $g \equiv (1+n)^j \times x \pmod{n^{t+1}}$ such that $\gcd(j,n) = 1$, $x \in \mathbb{Z}_n^*$ and

$$c_0 \equiv E_g(m_0, u_0) \equiv g^{m_0} \times (u_0)^{n^t} \pmod{n^{t+1}},$$
$$c_1 \equiv E_g(m_1, u_1) \equiv g^{m_1} \times (u_1)^{n^t} \pmod{n^{t+1}},$$

where $u_0, u_1 \in \mathbb{Z}_n^*$. Since $m_0, m_1 \in \mathbb{Z}_{n^t}$, A can write $m_0 = m_0' + k_0 \times n^{t-1}$ and $m_1 = m_1' + k_1 \times n^{t-1}$ such that $0 \le m_0', m_1' < n^{t-1}$ and $0 \le k_0, k_1 < n$. A calculates $g' \equiv g \equiv (1+n)^j \times x \pmod{n^t}$, which is a valid encryption base, and

$$\begin{aligned} c_0' &\equiv c_0 \pmod{n^t} \\ &\equiv g^{m_0} \times (u_0)^{n^t} \pmod{n^t} \\ &\equiv (g')^{m_0' + k_0 \times n^{t-1}} \times (u_0)^{n^t} \pmod{n^t}, \end{aligned}$$

and

$$\begin{aligned} c_1' &\equiv c_1 \pmod{n^t} \\ &\equiv g^{m_1} \times (u_1)^{n^t} \pmod{n^t} \\ &\equiv (g')^{m_1' + k_1 \times n^{t-1}} \times (u_1)^{n^t} \pmod{n^t}. \end{aligned}$$

Note that c_0' is a valid encryption of m_0' since

$$\begin{aligned} (c_0')^d &\equiv ((g')^{m_0' + k_0 \times n^{t-1}} \times (u_0)^{n^t})^d \pmod{n^t} \\ &\equiv ((1+n)^j \times x)^{(m_0' + k_0 \times n^{t-1}) \times d} \times (u_0)^{n^t \times d} \pmod{n^t} \\ &\equiv (1+n)^{j \times d \times (m_0' + k_0 \times n^{t-1}) \pmod{n^{t-1}}} \times \\ &\quad (x^{(m_0' + k_0 \times n^{t-1})} \times u_0^{n^t})^{d \pmod{\lambda(n)}} \pmod{n^t} \\ &\equiv (1+n)^{j \times d \times m_0'} \pmod{n^t}. \end{aligned}$$

This means that decryption of c_0' is m_0'. Similarly, c_1' is a valid encryption of m_1' since

$$\begin{aligned} (c_1')^d &\equiv ((g')^{m_1' + k_1 \times n^{t-1}} \times (u_1)^{n^t})^d \pmod{n^t} \\ &\equiv ((1+n)^j \times x)^{(m_1' + k_1 \times n^{t-1}) \times d} \times (u_1)^{n^t \times d} \pmod{n^t} \\ &\equiv (1+n)^{j \times d \times (m_1' + k_1 \times n^{t-1}) \pmod{n^{t-1}}} \times \\ &\quad \times (x^{(m_1' + k_1 \times n^{t-1})} \times u_1^{n^t})^{d \pmod{\lambda(n)}} \pmod{n^t} \\ &\equiv (1+n)^{j \times d \times m_1'} \pmod{n^t}. \end{aligned}$$

This means that the decryption of c_1' is m_1'. Thus, A can distinguish encryptions of m_0' and m_1' when $s = t$. This contradicts our assumption. Hence the semantic security of the Damgård-Jurik algorithm is equivalent to the security of D-Class$[n]$.

10.7 Example

Assume that the sender chooses $p = 13$, $q = 11$, $s = 3$ and calculates $n = 143$, $143^2 = 20449$, $143^3 = 2924207$, $143^4 = 418161601$. Then the sender selects $x = 1$, random $u = 7$ from $\mathbb{Z}_{143}^*$ and $j = 2$. So the semirandom base $g \in \mathbb{Z}_{143^4}^*$ becomes $g \equiv (1+n)^j \times x \equiv 144^2 \equiv 20736 \pmod{143^4}$. Hence the public key becomes $(n, g) = (143, 20736)$. To set the private decryption key, the sender calculates $\lambda(n) = \lambda(143) = \text{lcm}(\lambda(11), \lambda(13)) = \text{lcm}(10, 12) = 60$, then chooses either $d = 60$ or $d = 120 \in \mathbb{Z}_{143}^*$. In this example we assume that the private decryption key is $d = 120$. Let the message to be sent be $m = 10$ (modulo 143^3). Then the encrypted message is

$$\begin{aligned}
c &= E_{s,g}(m, u) \\
&= E_{3,20736}(10, 7) \\
&\equiv g^m \times u^{n^s} \pmod{n^{s+1}} \\
&\equiv 20736^{10} \times 7^{2924207} \pmod{143^4} \\
&\equiv 410352944 \times 196304565 \pmod{143^4} \\
&\equiv 36923343 \pmod{143^4}.
\end{aligned}$$

Once the encrypted message c arrives, the receiver computes c^d and g^d.

$$\begin{aligned}
c^d &\equiv 36923343^{120} \equiv 261967707 \pmod{143^4} \\
g^d &\equiv \quad 20736^{120} \quad \equiv 188819489 \pmod{143^4}.
\end{aligned}$$

Now the receiver needs to extract the powers of $c^d = (1+n)^{j \times d \times m} = (1+n)^i$ and $g^d = (1+n)^{j \times d} = (1+n)^h$ by using the discrete logarithm algorithm given by equivalence (10.5). First i_1 is extracted as follows:

$$\begin{aligned}
i_1 &\equiv L(261967707 \pmod{143^2}) \pmod{143} \\
&\equiv L(16017) \pmod{143} \\
&\equiv \frac{16016}{143} \pmod{143} \\
&\equiv 112 \pmod{143}.
\end{aligned}$$

Now using $i_1 = 112$, i_2 can be extracted as follows:

$$
\begin{aligned}
i_2 &\equiv L(261967707 \pmod{143^3}) - \binom{112}{2} \times 143 \pmod{143^2} \\
&\equiv L(1713284) - 56 \times 111 \times 143 \pmod{143^2} \\
&\equiv \frac{1713283}{143} - 888888 \pmod{143^2} \\
&\equiv 11981 - 888888 \pmod{143^2} \\
&\equiv 2400 \pmod{143^2}.
\end{aligned}
$$

Finally, $i_3 = i$ can be extracted as follows:

$$
\begin{aligned}
i_3 &\equiv L(261967707 \pmod{143^4}) - \binom{2400}{2} \times 143 - \binom{2400}{3} \times 143^2 \bmod 143^3 \\
&\equiv \frac{261967706}{143} - 1200 \times 2399 \times 143 - 400 \times 2399 \times 2398 \times 143^2 \bmod 143^3 \\
&\equiv 1831942 - 411668400 - 47055619239200 \pmod{143^3} \\
&\equiv 2400 \pmod{143^3}.
\end{aligned}
$$

Hence $c^d \equiv (1+n)^{2400}$. Now with the same method h can be extracted, starting from h_1:

$$
\begin{aligned}
h_1 &\equiv L(188819489 \pmod{143^2}) \pmod{143} \\
&\equiv L(13872) \pmod{143} \\
&\equiv \frac{13871}{143} \pmod{143} \\
&\equiv 97 \pmod{143}.
\end{aligned}
$$

Now using $h_1 = 97$, h_2 can be extracted as follows:

$$
\begin{aligned}
h_2 &= L(188819489 \pmod{143^3}) - \binom{97}{2} \times 143 \pmod{143^2} \\
&\equiv L(1670241) - 97 \times 48 \times 143 \pmod{143^2} \\
&\equiv \frac{1670240}{143} - 665808 \pmod{143^2} \\
&\equiv 11680 - 665808 \pmod{143^2} \\
&\equiv 240 \pmod{143^2}.
\end{aligned}
$$

Finally, $h_3 = h$ can be extracted as follows:

$$
\begin{aligned}
h_3 &\equiv L(188819489 \pmod{143^4}) - \binom{240}{2} \times 143 - \binom{240}{3} \times 143^2 \pmod{143^3} \\
&\equiv \frac{188819488}{143} - 120 \times 239 \times 143 - 40 \times 239 \times 238 \times 143^2 \pmod{143^3} \\
&\equiv 1320416 - 4101240 - 46527200720 \pmod{143^3} \\
&\equiv 240 \pmod{143^3}.
\end{aligned}
$$

Next, the receiver calculates

$$h^{-1} \equiv (j \times d)^{-1} \pmod{n^s} \equiv 240^{-1} \equiv 1181867 \pmod{143^3},$$

and then recovers the message m as follows:

$$
\begin{aligned}
m &\equiv i \times h^{-1} \pmod{n^s} \\
&\equiv (j \times m \times d) \times (j \times d)^{-1} \pmod{n^s} \\
&\equiv 2400 \times 1181867 \pmod{143^3} \\
&\equiv 10 \pmod{143^3}.
\end{aligned}
$$

Consider $m_1 = 10$, $u_1 = 7$. Then the encrypted message is $c_1 = 36923343$ as calculated before. To illustrate the additive homomorphic property of the algorithm, further let $m_2 = 9$, $u_2 = 5$. Then

$$
\begin{aligned}
c_2 &= E_{s,g}(m_2, u_2) \\
&= E_{3,20736}(9, 5) \\
&\equiv 20736^9 \times 5^{2924207} \pmod{143^4} \\
&\equiv 298476179 \times 17279166 \pmod{143^4} \\
&\equiv 339165159 \pmod{143^4}.
\end{aligned}
$$

If we compute $c' \equiv c_1 \times c_2 \pmod{143^4}$, we get

$$c' \equiv 36923343 \times 339165159 \equiv 345749719 \pmod{143^4}.$$

Now let $m_3 = m_1 + m_2 \equiv 19 \pmod{143^3}$, $u_3 \equiv u_1 \times u_2 \equiv 35 \pmod{143}$, and $c_3 = E_{s,g}(m_3, u_3)$. Hence we have

$$
\begin{aligned}
c_3 &= E_{s,g}(m_3, u_3) \\
&= E_{3,20736}(19, 35) \\
&\equiv 20736^{19} \times 35^{2924207} \pmod{143^4} \\
&\equiv 11456875 \times 287087544 \pmod{143^4} \\
&\equiv 345749719 \pmod{143^4}.
\end{aligned}
$$

We conclude that $c' \equiv c_3 \pmod{143^4}$, which confirms that $E_{s,g}(m_1 + m_2, u_1 \times u_2) = E_{s,g}(m_1, u_1) \times E_{s,g}(m_2, u_2)$, i.e. Eq. (10.6).

To illustrate that the algorithm supports the multiplication by a constant, consider the values $m = 10$, $u = 7$ and $c = 36923343$ from previous calculations. Let $k = 3$. Also let $m'' \equiv m \times k \equiv 30 \pmod{143^3}$ and $u'' \equiv u^k \equiv 7^3 \equiv 57 \pmod{143}$. Computing

$$c^k \equiv 36923343^3 \equiv 85484043 \pmod{143^4},$$

and also

$$\begin{aligned} c'' &= E_{s,g}(m'', u'') \\ &= E_{3,20736}(30, 57) \\ &\equiv 20736^{30} \times 57^{2924207} \pmod{143^4} \\ &\equiv 161944212 \times 372063908 \pmod{143^4} \\ &\equiv 85484043 \pmod{143^4}, \end{aligned}$$

we conclude $c^k \equiv c'' \pmod{143^4}$. Hence Eq. (10.7) is confirmed.

Next, to illustrate that the algorithm supports the addition by a constant, consider m, u and k. Let $c''' = E_{s,g}(m+k, u)$. Then we have

$$\begin{aligned} c''' &= E_{s,g}(m+k, u) \\ &= E_{3,20736}(13, 7) \\ &\equiv 20736^{13} \times 7^{2924207} \pmod{143^4} \\ &\equiv 82679026 \times 196304565 \pmod{143^4} \\ &\equiv 94593098 \pmod{143^4}. \end{aligned}$$

Computing $g^k \times c \equiv 20736^3 \times 36923343 \equiv 94593098 \pmod{143^4}$, we see that $E_{s,g}(m+k, u) = g^k \times E_{s,g}(m, u)$, which confirms Eq. (10.8). Consider $u' = 5$ for re-randomization of $c = E_{s,g}(m, u)$. Note that $u \times u' \equiv 7 \times 5 \equiv 35 \pmod{143}$, and calculating

$$\begin{aligned} c' &\equiv (u')^{n^s} \times c \pmod{143^4} \\ &\equiv 5^{143^3} \times E_{3,20736}(10, 7) \pmod{143^4} \\ &\equiv 17279166 \times 36923343 \pmod{143^4} \\ &\equiv 364508602 \pmod{143^4}, \end{aligned}$$

$$\begin{aligned} E_{3,20736}(10, 7 \times 5) &\equiv E_{3,20736}(10, 35) \pmod{143^4} \\ &\equiv 20736^{10} \times 35^{143^3} \pmod{143^4} \\ &\equiv 410352944 \times 287087544 \pmod{143^4} \\ &\equiv 364508602 \pmod{143^4}, \end{aligned}$$

we see that $D(5^{143^3} \times E_{3,20736}(10, 7)) = D(E_{3,20736}(10, 35))$, which confirms Eq. (10.9). Decrypting the re-randomized ciphertext $c' = 364508602$, we recover $m = 10$ by previous calculations.

Chapter 11
Boneh-Goh-Nissim Algorithm

The Boneh-Goh-Nissim algorithm is a public-key cryptosystem proposed by Boneh, Goh and Nissim in 2005 [4]. It is an additive somewhat homomorphic algorithm, which allows multiplication of text messages only once. The security of this algorithm relies on the problem of deciding whether an element x of a cyclic group G belongs to a specific subgroup of G, where $|G| = n$ and $n = q_1 \times q_2$, without knowing the factorization of n, which is known as the Subgroup Decision Problem. Here q_1 and q_2 are chosen to be distinct large primes.

Bilinear Pairings

We need to use a bilinear pairing to perform single multiplication of messages.

Definition 2 *Let G_1, G_2 be additive finite abelian groups and let G_3 be a multiplicative finite abelian group. A bilinear pairing is a map $e : G_1 \times G_2 \longrightarrow G_3$ for which the following hold:*

(i) (Bilinearity) For any $P, Q \in G_1$ and $R \in G_2$ we have

$$e(P+Q,R) = e(P,R) \times e(Q,R).$$

Similarly, for any $P \in G_1$ and $Q, R \in G_2$ we have

$$e(P,Q+R) = e(P,Q) \times e(P,R).$$

(ii) (Strong non-degeneracy) For any $P \in G_1$ with $P \neq 1_{G_1}$ there is some $Q \in G_2$ such that $e(P,Q) \neq 1_{G_3}$, and for any $Q \in G_2$ with $Q \neq 1_{G_2}$ there is some $P \in G_1$ such that $e(P,Q) \neq 1_{G_3}$.

In the case $G_1 = G_2 = G$ we say e is symmetric. Namely, we have $e(P,Q) = e(Q,P)$, since for a generator $R \in G$, there exists integers u, v such that $P = uR$ and $Q = vR$. Hence

$$e(P,Q) = e(uR,vR) = e(R,R)^{u \times v} = e(vR,uR) = e(Q,P).$$

Ç. K. Koç et al., *Partially Homomorphic Encryption*,
https://doi.org/10.1007/978-3-030-87629-6_11

There are a few types of pairings that are used for cryptographic purposes. The Weil pairing, which was introduced by André Weil [28], has been used to convert a discrete logarithm problem on elliptic curves (see Chap. 3 of [12] for the definition and properties of the Weil pairing). Although it was used to attack, it can be used for encryption with a slight modification. The modified Weil pairing, which is the Weil pairing equipped with a distortion map, serves well for cryptographic purposes instead of the Weil pairing. Since the Weil pairing is not strongly non-degenerate, it is modified to be strongly non-degenerate by using a distortion map. For more information on distortion maps and modifications see Chap. 3 of [13] and Chap. 6 of [12]. For practical use of bilinear pairings in cryptography, the pairing e should be computable in polynomial time. For efficient calculation algorithms of the modified Weil pairing we refer the reader to Chap. 3 of [11] and [16].

Construction of Bilinear Groups of Order *n*

Boneh, Goh and Nissim described the following method to construct a bilinear group G of a given order n, which is a group supporting a bilinear map [4]. Let $n > 3$ be a square-free integer that is not divisible by 3.

(i) Find the smallest positive integer $l \in \mathbb{Z}$ such that $p = l \times n - 1$ is prime and $p \equiv 2 \pmod 3$.
(ii) Consider the group of points on the elliptic curve $y^2 = x^3 + 1$ defined over $\mathbb{Z}_p$. Since $p \equiv 2 \pmod 3$, the curve has $p + 1 = l \times n$ points in $\mathbb{Z}_p \times \mathbb{Z}_p$. Therefore the group of points on the curve has a subgroup of order n, which is denoted by G.
(iii) Let G' be the subgroup of $\mathbb{Z}^*_{p^2}$ of order n. The modified Weil pairing on the curve gives a bilinear map $e : G \times G \longrightarrow G'$ with the required properties.

Although Boneh et al. have used the curve $y^2 = x^3 + 1$ when $p \equiv 2 \pmod 3$, the following proposition gives two different supersingular curves for construction of bilinear subgroups (see [27], p. 126).

Proposition 6 *Let p be a prime with $p \nmid n$ and $p \geq 5$. Then the elliptic curve $y^2 = x^3 + 1$ over $\mathbb{Z}_p$ is supersingular if and only if $p \equiv 2 \pmod 3$, and the elliptic curve $y^2 = x^3 + x$ over $\mathbb{Z}_p$ is supersingular if and only if $p \equiv 3 \pmod 4$.*

An advantage of working with supersingular curves is that the number of points on the curve can be found easily.

Proposition 7 *Suppose $p \geq 5$ is a prime. Then $\mathscr{E}$ is supersingular if and only if $\#\mathscr{E}(\mathbb{Z}_p) = p + 1$, where $\#\mathscr{E}(\mathbb{Z}_p)$ denotes the number of points in $\mathbb{Z}_p \times \mathbb{Z}_p$ lying on the curve.*

Remark 1 *Note that choosing a prime p such that $p + 1 = l \times n$ and using a supersingular curve $\mathscr{E}$ enables us to construct a group of order n, since $\mathscr{E}(\mathbb{Z}_p)$ is a group of order $p + 1$, which is a multiple of n.*

11.1 Key Generation

From now on, for convenience, we drop $\oplus$, use the usual multiplication sign "$\times$" instead, and refer to $[a]P$ as P^a. In other words, P^a refers to the result of applying the group operation of G successively a times on P. Indeed,

$$P^a = P \oplus P \oplus \cdots \oplus P = [a]P$$

on an elliptic curve.

Key generation involves the following steps:

(i) Choose two large primes q_1 and q_2 and compute $n = q_1 \times q_2$.
(ii) Choose a larger prime p such that $p = l \times n - 1$ for some positive integer l.
(iii) Find a cyclic group G which is a subgroup of $\mathscr{E}(\mathbb{Z}_p)$ of order $n = q_1 \times q_2$ and a bilinear pairing $e : G \times G \longrightarrow G'$, where G' is a subgroup of $\mathbb{Z}^*_{p^2}$.
(iv) Choose two distinct random generators g, u of G and also set $h = u^{q_2}$. Then h is a random generator of the order q_1 subgroup of G.

The public encryption key (n, G, G', e, g, h) is released and the private decryption key q_1 is kept secret.

Algorithm 39 Algorithm to find the order of a point in an elliptic curve group $\mathscr{E}(\mathbb{Z}_p)$

```
1: function PORDER(𝓔,p,g)
2:     Input: Elliptic curve 𝓔, a prime p, a point g in 𝓔(Z_p).
3:     Output: Order of the point g in 𝓔(Z_p).
4:     S:=ECURVE(𝓔,p) a               ▷ Gives the elliptic curve group 𝓔(Z_p).
5:     i:=0                            ▷ Initializes an element.
6:     g':=𝒪                           ▷ Initializes an element to the point at infinity.
7:     repeat
8:         g':=PADD(S,g',g) b           ▷ Adds the points g,g' in S.
9:         i:=i+1
10:    until g' = 𝒪                    ▷ Calculates order of the point g in 𝓔(Z_p).
11:        return i
```

[a] This function creates the elliptic curve group $\mathscr{E}(\mathbb{Z}_p)$ with inputs $\mathscr{E}$ and p.
[b] Given S and points $g, g' \in S$, this function performs point addition on an elliptic curve group S.

11.2 Encryption

After converting an alphanumeric plaintext message M into a purely numeric message $m \in \mathbb{Z}_{q_2}$, the sender chooses a random $r \in \mathbb{Z}_n$. Then the encrypted message c is computed as

Algorithm 40 Key generation algorithm for the Boneh-Goh-Nissim cryptosystem

1: **Input:** Large odd distinct k-bit primes q_1, q_2.
2: **Output:** Public key (n, G, g, h) and secret key q_1.
3: $n := q_1 \times q_2$
4: **repeat** ▷ Chooses p.
5: l:=RANDINT$(1, 2^{50} - 1)$
6: i:= $(n \times l - 1) \pmod 4$
7: j:=$(n \times l - 1) \pmod 3$
8: **until** ISPRIME$(n \times l - 1)$ is true and $(i = 3$ or $j = 2)$ is true.
9: p:=$n \times l - 1$
10: **if** $i = 3$ and $j \neq 2$ **then**
11: $\mathscr{E} := x^3 + x$
12: **if** $i \neq 3$ and $j = 2$ **then**
13: $\mathscr{E} := x^3 + 1$
14: **else** $\mathscr{E} := x^3 + 1$
▷ This is when both $i = 3$ and $j = 2$ are true. User might also choose $\mathscr{E} := x^3 + x$ as well.
15: S:=ECURVE$(\mathscr{E}, p)$ ▷ Gives the elliptic curve group $\mathscr{E}(\mathbb{Z}_p)$.
16: **repeat**
17: g:=RANDP(S)[a] ▷ Chooses encryption base.
18: **until** PORDER$(\mathscr{E}, p, g)$=n
19: G:=GEN$(\mathscr{E}, p, g)$[b] ▷ Generates G.
20: **repeat**
21: r:=RANDINT(2,n-1)
22: **until** GCD(r, n)=1
23: u:=PDOUB(S, g, r)[c] ▷ Chooses u.
24: h:=PDOUB(S, u, q_2)
25: **return** (n, G, g, h) and q_1

[a] Given an elliptic curve group S, this function chooses a random point $g \in S$.
[b] This function gives the subgroup of $\mathscr{E}(\mathbb{Z}_p)$ generated by a point $g \in \mathscr{E}(\mathbb{Z}_p)$ with inputs $\mathscr{E}, p, g$.
[c] Given an elliptic curve group S, a point $g \in S$ and an integer r, this function performs point multiplication and calculates g^r on S.

$$c = g^m \times h^r \in G. \tag{11.1}$$

A different r-value is assigned to a message each time.

Algorithm 41 Encryption algorithm for the Boneh-Goh-Nissim cryptosystem

1: **Input:** Message m, encryption key (n, G, g, h).
2: **Output:** Encrypted message c.
3: r:=RANDINT(0,n-1) ▷ Chooses randomness.
4: c:=PADD(G,PDOUB(G, g, m),PDOUB(G, h, r))
5: **return** c

11.3 Decryption

Once the receiver gets the encrypted message c, the original message m can be recovered using the private decryption key q_1. First note that

$$\begin{aligned} c^{q_1} &= (g^m \times h^r)^{q_1} \\ &= (g^m)^{q_1} \times (h^r)^{q_1} \\ &= (g^m)^{q_1} \times ((u^{q_2})^r)^{q_1} \\ &= (g^{q_1})^m \times (u^{q_1 \times q_2})^r \\ &= (g^{q_1})^m \in G \end{aligned}$$

as $\mathrm{ord}(u) = n = q_1 \times q_2$ in G. This means, to recover m, it suffices to compute the discrete logarithm of c^{q_1} base g^{q_1} using Pollard's lambda method [20].

Algorithm 42 Decryption algorithm for the Boneh-Goh-Nissim cryptosystem

1: **Input:** Received message c, elliptic curve group G, base of encryption g, decryption key q_1.
2: **Output:** Original message m.
3: c':=PDOUB(G, c, q_1)
4: m:=LAMBDA(G, c', g^{q_1})[a]
5: **return** m

[a] Given an elliptic curve group G, a point $c' \in G$ and a base point (which is g^{q_1} in this case), this function uses the lambda algorithm to find the logarithm of c' base g^{q_1}.

11.4 Homomorphic Properties

Let $n = q_1 \times q_2$ and $g, u \in G^*$ such that $u^{q_2} = h$. Then the encryption function can be described as:

$$\begin{aligned} E_g : \mathbb{Z}_{q_2} \times \mathbb{Z}_n &\to G \\ (m, r) &\mapsto g^m \times h^r, \end{aligned}$$

where $r \in \mathbb{Z}_n$ is random and $\times$ denotes the group operation of G. Note that the encryption function is very similar to that of the Okamoto-Uchiyama algorithm. Hence all homomorphic properties of the Okamoto-Uchiyama algorithm are inherited by BGN. Recall that the sender changes the randomness $r \in \mathbb{Z}_n$ for every message to be encrypted. Denote the decryption function by D. Consider the messages m_1 and m_2 with their accompanying random r-values r_1 and r_2, respectively. Then the BGN algorithm is considered to be additively homomorphic with the accompanying r-value of the message $m_1 + m_2$ being $r_1 + r_2$.

$$\begin{aligned} E_g(m_1,r_1) \times E_g(m_2,r_2) &= (g^{m_1} \times h^{r_1}) \times (g^{m_2} \times h^{r_2}) \\ &= g^{m_1+m_2} \times h^{r_1+r_2} \\ &= E_g(m_1+m_2, r_1+r_2), \end{aligned}$$

where the randomness of the message m_1+m_2 is r_1+r_2, which is neither uniformly distributed in G nor independent of the randomness in $E_g(m_1,r_1)$ and $E_g(m_2,r_2)$. However this can be addressed by re-randomization, which is explained at the end of this section. This means that if we denote $E_g(m,r) = E(m)$, then for two messages m_1 and m_2 we have

$$D(E(m_1+m_2)) = D(E(m_1) \times E(m_2)), \tag{11.2}$$

where D denotes the decryption function.

For a message m with randomness r and a scalar k, scalar multiplication works for this algorithm by fixing the randomness of $k \times m$ to $k \times r$.

$$\begin{aligned} (E_g(m,r))^k &= (g^m \times h^r)^k \\ &= g^{k \times m} \times h^{k \times r} \\ &= E_g(k \times m, k \times r), \end{aligned}$$

which means

$$D(E(k \times m)) = D((E(m))^k). \tag{11.3}$$

The BGN algorithm allows a single multiplication of plaintexts by changing the base and the range of the encryption function once. Since g and u are generators of G and $h = u^{q_2}$, let $g^\alpha = h$ for some $\alpha < n$. Suppose that $e : G \times G \to G'$ is a bilinear pairing and $e(g,g) = g'$, $e(g,h) = h'$. Hence

$$h' = e(g,h) = e(g,g^\alpha) = e(g,g)^\alpha = (g')^\alpha.$$

Note that g being a generator of G implies that $e(g,g) = g'$ is a generator of G' and that G' is a group of order n. Then

$$\begin{aligned} E_{g'} : \mathbb{Z}_{q_2} \times \mathbb{Z}_n &\to G' \\ (m,r) &\mapsto (g')^m \times (h')^r, \end{aligned}$$

where $r \in \mathbb{Z}_n$ is random and $\times$ denotes the group operation of G'. We simply pair two encrypted messages $c_1 = E_g(m_1,r_1)$ and $c_2 = E_g(m_2,r_2)$ with base g to get a valid encryption of $m_1 \times m_2$ with base g':

$$\begin{aligned}
e(c_1,c_2)\times(h')^r &= e(g^{m_1}\times h^{r_1}, g^{m_2}\times h^{r_2})\times(h')^r \\
&= e(g^{m_1}\times(g^{\alpha})^{r_1}, g^{m_2}\times(g^{\alpha})^{r_2})\times(h')^r \\
&= e(g^{m_1}\times g^{\alpha\times r_1}, g^{m_2}g^{\alpha\times r_2})\times(h')^r \\
&= e(g^{m_1+\alpha\times r_1}, g^{m_2+\alpha\times r_2})\times(h')^r \\
&= e(g,g)^{(m_1+\alpha\times r_1)\times(m_2+\alpha\times r_2)}\times(h')^r \\
&= (g')^{(m_1+\alpha\times r_1)\times(m_2+\alpha\times r_2)}\times(h')^r \\
&= (g')^{m_1\times m_2+\alpha\times m_1\times r_2+\alpha\times m_2\times r_1+\alpha^2\times r_1\times r_2}\times(h')^r \\
&= (g')^{m_1\times m_2+\alpha\times(m_1\times r_2+m_2\times r_1+\alpha\times r_1\times r_2)}\times(h')^r \\
&= (g')^{m_1\times m_2}\times(g')^{\alpha\times(m_1\times r_2+m_2\times r_1+\alpha\times r_1\times r_2)}\times(h')^r \\
&= (g')^{m_1\times m_2}\times((g')^{\alpha})^{m_1\times r_2+m_2\times r_1+\alpha\times r_1\times r_2}\times(h')^r \\
&= (g')^{m_1\times m_2}\times(h')^{m_1\times r_2+m_2\times r_1+\alpha\times r_1\times r_2}\times(h')^r \\
&= (g')^{m_1\times m_2}\times(h')^{m_1\times r_2+m_2\times r_1+\alpha\times r_1\times r_2+r} \\
&= (g')^{m_1\times m_2}\times(h')^{\overline{r}} \\
&= E_{g'}(m_1\times m_2,\overline{r}).
\end{aligned}$$

The term $E_{g'}(m_1 \times m_2, \overline{r})$ is a valid encryption of $m_1 \times m_2$. Also, $E_{g'}$ inherits all the homomorphic properties of E_g, except single multiplication of text messages. This multiplication can be done only once, since there is no known bilinear pairing e' from $G' \times G'$ to any other group of order n.

On the other hand, no multiplicatively homomorphic operation $\boxtimes$ is known such that

$$D(E(m_1\times m_2)) = D(E(m_1)\boxtimes E(m_2))$$

for the BGN algorithm, which can be applied infinitely many times. Therefore BGN is not considered to be multiplicatively homomorphic. With the help of the bilinear pairing e and the additive homomorphic properties of BGN given by Eq. (11.2), we can perform addition as often as we want before and after a single multiplication of plaintext messages.

Another homomorphic property supported by the BGN algorithm is addition with a constant.

$$\begin{aligned}
E_g(m+k,r) &= g^{m+k}\times(h)^r \\
&= (g^m\times g^k)\times h^r \\
&= (g^m\times h^r)\times g^k \\
&= c\times g^k,
\end{aligned}$$

which means

$$D(E(k+m)) = D(E(m)\times g^k). \tag{11.4}$$

Re-randomizaton of messages is also possible for the BGN algorithm. Let $c = E_g(m,r)$ and $r' \in \mathbb{Z}_n$, then

$$c \times h^{r'} = g^m \times h^r \times h^{r'} = g^m \times h^{r+r'}$$

is a valid encryption of m, since $r + r' \in \mathbb{Z}_n$. Therefore

$$D(c \times h^{r'}) = D(c). \tag{11.5}$$

11.5 Security

The semantic security of the BGN algorithm relies on the Subgroup Decision Problem.

Theorem 44 *The BGN algorithm is semantically secure if the Subgroup Decision Problem is intractable.*

Proof. We prove the contrapositive statement, which is logically equivalent. Assume that the Subgroup Decision Problem is not intractable. Then an adversary A can decide whether $c = g^m \times h^r \in G$ belongs to a proper subgroup of G, where $h = u^{q_2}$ and $r \in \mathbb{Z}_n$. Given that c is a valid encryption of either m_0 or m_1 ($m_0 < m_1$), A calculates $c \times g^{-m_0} = g^{m-m_0} \times h^r$. If c is the encryption of m_0 then

$$\begin{aligned}
(c \times g^{-m_0})^{q_1} &= (h^r)^{q_1} \\
&= (u^{q_2})^{r \times q_1} \\
&= u^{r \times q_1 \times q_2} \\
&= (u^n)^r \\
&= 1_G,
\end{aligned}$$

which means that $c = g^m \times h^r$ belongs to the subgroup of G of order q_1. If c is the encryption of m_1 then

$$\begin{aligned}
(c \times g^{-m_0})^{q_1} &= (g^{(m_1-m_0)} \times h^r)^{q_1} \\
&= (g^{(m_1-m_0)})^{q_1} \times (u^{q_2})^{r \times q_1} \\
&= g^{(m_1-m_0) \times q_1} \times u^{r \times q_1 \times q_2} \\
&= g^{(m_1-m_0) \times q_1} \times (u^n)^r \\
&= g^{(m_1-m_0) \times q_1} \\
&\neq 1_G,
\end{aligned}$$

since $0 < m_1 - m_0 < q_2$ and $\mathrm{ord}(g) = n$ in G. Hence $c = g^m \times h^r$ does not belong to the subgroup of G of order q_1. Also

$$\begin{aligned}
(c \times g^{-m_0})^{q_2} &= (g^{(m_1-m_0)} \times h^r)^{q_2} \\
&= (g^{(m_1-m_0)})^{q_2} \times (u^{q_2})^{r \times q_2} \\
&= g^{(m_1-m_0) \times q_2} \times u^{r \times q_2 \times q_2},
\end{aligned}$$

so $c \times g^{-m_0}$ does not belong to the subgroup of G of order q_2 with overwhelming probability. Therefore A decides that c is an encryption of m_0 if $c \times g^{-m_0}$ belongs to a proper subgroup of G, and it is an encryption of m_1 otherwise. Since A can distinguish encrypted messages, the BGN algorithm is not semantically secure.

Note that if semantic security holds for ciphertexts in G then it also holds for ciphertexts in G'. Assume that semantic security does not hold for ciphertexts in G'. Then adversary A can distinguish $c = g^m \times h^r \in G$ by calculating

$$\begin{aligned} e(c, c_1) \times (h')^r &= E_{g'}(m \times 1, \bar{r}) \\ &= E_{g'}(m, \bar{r}), \end{aligned}$$

where c_1 is a valid encryption of 1.

11.6 Example

We start with key generation, i.e. construction of the cyclic group. Let $q_1 = 7$, $q_2 = 11$ and compute $n = q_1 \times q_2 = 77$. Next we construct an elliptic curve group (cyclic subgroup) of order $n = 77$ that has an associated bilinear map e. First note that $p = 4 \times 77 - 1 = 307$ is prime and $p \equiv 3 \pmod 4$. By Proposition 7 and Proposition 6, $\mathscr{E} : y^2 = x^3 + x$ over $\mathbb{Z}_{307}$ is supersingular with $\#\mathscr{E}(\mathbb{Z}_{307}) = 307 + 1 = 308$, which contains a subgroup G of order $308/4 = 77 = n$. $g = (182, 240)$ is a point of order 77. $u = (182, 240)^{48} = (28, 262)$ and u is also of order 77. So u and g are two different generators of the same cyclic group of order 77 on the elliptic curve. Then we compute $h = u^{q_2} = (28, 262)^{11} = (99, 120)$, where h is of order $q_1 = 7$. Since we deal with small numbers, we can construct a table to calculate discrete logarithms with base g^{q_1}.

$$\begin{aligned} (g^{q_1})^1 &= (146, 60) & (g^{q_1})^2 &= (299, 44) \\ (g^{q_1})^3 &= (272, 206) & (g^{q_1})^4 &= (191, 151) \\ (g^{q_1})^5 &= (79, 171) & (g^{q_1})^6 &= (79, 136) \\ (g^{q_1})^7 &= (191, 156) & (g^{q_1})^8 &= (272, 101) \\ (g^{q_1})^9 &= (299, 263) & (g^{q_1})^{10} &= (146, 247) \\ (g^{q_1})^{11} &= \mathscr{O} \end{aligned}$$

Let $m = 3$. Take $r = 2$ and compute $c = E_g(m, r)$.

$$c = g^m \times h^r = (182, 240)^3 \times (99, 120)^2 = (287, 283) \times (175, 229) = (177, 88).$$

To decrypt the ciphertext c, we compute the discrete logarithm of c^{q_1} with base g^{q_1}. First compute

$$c^{q_1} = (177, 88)^7 = (272, 206).$$

Note that $(g^{q_1})^3 = (272, 206) = c^{q_1}$. Therefore we recover $m = 3$.

Let $m_1 = 3$, $m_2 = 2$, $r_1 = 2$ and $r_2 = 3$. We have $c_1 = E_g(m_1, r_1) = (177, 88)$ from previous calculations. We compute $c_2 = E_g(m_2, r_2)$ as follows:

$$c_2 = g^{m_2} \times h^{r_2} = (182, 240)^2 \times (99, 120)^3 = (259, 143) \times (40, 106) = (277, 39).$$

Then we compute

$$E_g(m_1, r_1) \times E_g(m_2, r_2) = (177, 88) \times (277, 39) = (294, 218).$$

On the other hand,

$$E_g(m_1 + m_2, r_1 + r_2) = E_g(5, 5) = (182, 240)^5 \times (99, 120)^5 = (294, 218),$$

which confirms Eq. (11.2).

Let $m' = 3$, $r' = 5$, and choose scalar $k' = 2$. Compute $(c')^{k'}$, where $c' = E_g(m', r')$.

$$(c')^{k'} = ((182, 240)^3 \times (99, 120)^5)^2 = ((287, 283) \times (175, 78))^2 = (10, 150)^2 = (294, 89).$$

On the other hand,

$$\begin{aligned} E_g(k' \times m', k' \times r') &= E_g(6, 10) \\ &= g^6 \times h^{10} \\ &= (182, 240)^6 \times (99, 120)^{10} \\ &= (219, 139) \times (40, 106) \\ &= (294, 89), \end{aligned}$$

which confirms Eq. (11.3).

Let $m'' = 3$, $r'' = 2$ and $k'' = 3$, a constant. Calculate

$$\begin{aligned} E_g(m'' + k'', r'') &= g^{m'' + k''} \times h^{r''} \\ &= (182, 240)^6 \times (99, 120)^2 \\ &= (219, 139) \times (175, 229) \\ &= (169, 289). \end{aligned}$$

On the other hand, since $E_g(m'', r'') = (177, 88)$ by previous calculations,

$$g^{k''} \times E_g(m'', r'') = (182, 240)^3 \times (177, 88) = (169, 289),$$

which confirms Eq. (11.4).

Consider again $m = 3$ and $r = 2$. Choose $r' = 3$ for re-randomization. Note that $r + r' \equiv 2 + 3 \equiv 5 \pmod{77}$. Calculating

$$
\begin{aligned}
c \times h^{r'} &= E_g(m,r) \times h^{r'} \\
&= (177,88) \times (99,120)^3 \\
&= (177,88) \times (40,106) \\
&= (10,150),
\end{aligned}
$$

and also

$$
\begin{aligned}
E_g(m,r+r') &= E_g(3,5) \\
&= (182,240)^3 \times (99,120)^5 \\
&= (287,283) \times (175,78) \\
&= (10,150),
\end{aligned}
$$

we see that

$$
D((10,150)) = D(c \times h^{r'}) = D(E_g(m,r+r')) = m,
$$

which confirms Eq. (11.5).

Chapter 12
Sander-Young-Yung Algorithm

The Sander-Young-Yung cryptosystem was introduced by Tomas Sander, Adam Young and Moti Yung in 1999 [24]. In this scheme encryptions are not of the messages directly but of the encodings of messages. This scheme is homomorphic on $(\mathbb{Z}_2, \times)$ (i.e. allows one to compute the AND on encrypted values) and utilizes the Goldwasser-Micali scheme during encryption of encodings due to its additively homomorphic and efficient re-randomization properties. The security and semantic security of the SYY cryptosystem follow directly from the security and semantic security of the GM cryptosystem, which are based on the Quadratic Residuosity Problem (QR$[n]$) modulo $n = p \times q$ (see Sect. 2.1.10).

12.1 Key Generation

For key generation:

(i) Given a security parameter, choose two random large primes p and q, and compute $n = p \times q$, which is the size of the security parameter.
(ii) Choose a positive nonzero integer l.
(iii) Choose a quadratic non-residue $x \in \mathbb{Z}_n^*$ with Jacobi symbol $\left(\frac{x}{n}\right) = 1$.

Then the public key is (n, x, l) and (p, q) is the private key.

12.2 Encryption

The messages are chosen from $\mathbb{Z}_2$. Before encryption of messages, they are encoded via the following map

$$\mathscr{E} : \mathbb{Z}_2 \to \mathbb{Z}_2^l,$$

Ç. K. Koç et al., *Partially Homomorphic Encryption*,
https://doi.org/10.1007/978-3-030-87629-6_12

Algorithm 43 Key generation algorithm for the Sander-Young-Yung cryptosystem

1: **Input:** Large random primes p,q and a positive integer l.
2: **Output:** Public key (n,x,l) and private key (p,q).
3: $n := p \times q$
4: **repeat**
5: x:=RANDINT$(1,n-1)$
6: **until** GCD(x,n)=1 and JAC(x,p)=-1 and JAC(x,q)=-1
▷ Choose a quadratic non-residue x in $\mathbb{Z}_n^*$ with Jacobi symbol $(x/n) = 1$.
7: **return** (n,x,l) and (p,q)

which encodes 1 by the zero vector in $\mathbb{Z}_2^l$ and encodes 0 by a random nonzero vector in $\mathbb{Z}_2^l$; and decoded via

$$\mathscr{D} : \mathbb{Z}_2^l \to \mathbb{Z}_2,$$

which decodes any nonzero vector to 0 and decodes the zerovector to 1.

The AND-operation (or multiplication) on $\mathbb{Z}_2$ is executed by the addition of the vector encodings (i.e. the componentwise XOR of the vector encodings). If both vectors are the zero vector, i.e. they are encodings of 1, then the resulting vector is also the zero vector, which decodes to 1. Formally, $\mathscr{D}(\mathscr{E}(1 \times 1)) = \mathscr{D}(\mathscr{E}(1) + \mathscr{E}(1)) = 1$. If one of the factors is not the zero vector, then the resulting vector is also nonzero, which decodes to 0. Formally, $\mathscr{D}(\mathscr{E}(1 \times 0)) = \mathscr{D}(\mathscr{E}(1) + \mathscr{E}(0)) = 0$. If both vectors are nonzero vectors, then the resulting vector is nonzero except in the case of probability $1/2^l$ that the two vectors are equal, when it decodes to 1. Formally, $\mathscr{D}(\mathscr{E}(0 \times 0)) = \mathscr{D}(\mathscr{E}(0) + \mathscr{E}(0)) = 0$ with probability $1 - 1/2^l$. Note that to prevent getting the same encoding of 0 simultaneously, re-randomization through matrices is possible, which is described in the following lemma.

Lemma 20 *Let A be chosen uniformly at random to be a non-singular $l \times l$ matrix over $\mathbb{Z}_2$ and $v \in \mathbb{Z}_2^l$ be a nonzero vector. Then $Av \in \mathbb{Z}_2^l$ is also a nonzero vector.*

After encoding of message $m \in \mathbb{Z}_2$ to a vector $(v_1, \ldots, v_l) \in \mathbb{Z}_2^l$, this vector is encrypted by using the Goldwasser-Micali encryption function E as follows:

$$EV(m) := E\big(\mathscr{E}(m)\big) = E(v_1, \ldots, v_l) = \big(E(v_1), \ldots, E(v_l)\big) = ({y_1}^2 \times x^{v_1}, \ldots, {y_l}^2 \times x^{v_l}),$$

where $y_i \in \mathbb{Z}_n^*$ are chosen uniformly at random.

12.3 Decryption

The receiver, who holds the private key (p,q), can decrypt the ciphertext tuple by deciding the quadratic residuosity of c_i modulo p and modulo q by computing the Legendre symbols. If c_i is a quadratic residue modulo both p and q, then c_i is a quadratic residue modulo n by Theorem 18. If c_i is a quadratic residue modulo n, then $v_i = 0$; if c_i is a quadratic non-residue modulo n, then $v_i = 1$. Formally,

Algorithm 44 Encryption algorithm for the Sander-Young-Yung cryptosystem

1: **Input:** One-bit message m and encryption key (n,x,l).
2: **Output:** Ciphertext string $(c_1,c_2,\ldots,c_l)$.
3: **while** $m=1$ **do**
4:     **for** $i=1$ **to** l **do** $v_i:=0$
5:     $\mathscr{E}:=(v_1,v_2,\ldots,v_l)$ ▷ 1 is encoded as a zero vector in $\mathbb{Z}_2^l$.
6:     **for** $i=1$ **to** l **do**
7:         **repeat**
8:             $y_i:=$ RANDINT$(1,n-1)$
9:         **until** GCD(y_i,n)=1
10:         $c_i:=y_i^2 \pmod n$
11: **while** $m=0$ **do**
12:     **repeat**
13:         **for** $i=1$ **to** l **do**
14:             $v_i:=$ RANDINT$(0,1)$
15:     **until** $(v_1,\ldots,v_l)\neq(0,\ldots 0)$
16:     $\mathscr{E}:=(v_1,\ldots,v_l)$ ▷ 0 is encoded as a nonzero vector in $\mathbb{Z}_2^l$.
17:     **for** $i=1$ **to** l **do**
18:         **repeat**
19:             $y_i:=$ RANDINT$(1,n-1)$
20:         **until** GCD(y_i,n)=1
21:         $c_i:=y_i^2\times x^{v_i} \pmod n$
22: **return** $(c_1,c_2,\ldots,c_l)$

$$D\big(E(v_1),\ldots,E(v_l)\big)=\big(D(E(v_1)),\ldots,D(E(v_l))\big)=(v_1,\ldots v_l),$$

where D is the decryption function of the Goldwasser-Micali scheme. If the obtained vector is the zero vector, then the message decodes to 1, else to 0.

Algorithm 45 Decryption algorithm for the Sander-Young-Yung cryptosystem

1: **Input:** Received message string $(c_1,c_2,\ldots,c_l)$, decryption key (p,q).
2: **Output:** Original message m.
3: **for** $i=1$ **to** l **do**
4:     **if** JAC(c_i,p)=1 and JAC(c_i,q)=1 **then** $v_i:=0$
5:     **else** $v_i:=1$
6: **if** $(v_1,\ldots,v_l)=(0,\ldots,0)$ **then** $m:=1$
7: **else** $m:=0$
8: **return** m

12.4 Homomorphic Properties

The encryption function of the Sander-Young-Yung algorithm is

$$
\begin{aligned}
E : \mathbb{Z}_2 &\to (\mathbb{Z}_n^*)^l \\
m &\mapsto ({y_1}^2 \times x^{v_1} \pmod{n}, \ldots, {y_l}^2 \times x^{v_l} \pmod{n}),
\end{aligned}
$$

where $(v_1, \ldots, v_l)$ is the encoding of m. Let m_0 and m_1 be encoded as $v = (v_1, \ldots, v_l)$ and $w = (w_1, \ldots, w_l)$, respectively. Further let $A = (a_{ij})$ and $B = (b_{ij})$ be non-singular $l \times l$ matrices chosen uniformly at random. Then the AND-homomorphic (or multiplicative) property $EV(m_0 \times m_1)$ is shown as

$$
\begin{aligned}
&= E\big(\mathscr{E}(m_0 \times m_1)\big) \\
&= E\big(\mathscr{E}(m_0) + \mathscr{E}(m_1)\big) \\
&= E(v + w) \\
&= E(Av + Bw) \\
&= E\Big(\sum_{j, a_{1j}=1} v_j \oplus \sum_{j, b_{1j}=1} w_j, \ldots, \sum_{j, a_{lj}=1} v_j \oplus \sum_{j, b_{lj}=1} w_j\Big) \\
&= \Big(E\Big(\sum_{j, a_{1j}=1} v_j \oplus \sum_{j, b_{1j}=1} w_j\Big), \ldots, E\Big(\sum_{j, a_{lj}=1} v_j \oplus \sum_{j, b_{lj}=1} w_j\Big)\Big) \\
&\stackrel{\star}{=} \Big(E\Big(\sum_{j, a_{1j}=1} v_j\Big) \times E\Big(\sum_{j, b_{1j}=1} w_j\Big), \ldots, E\Big(\sum_{j, a_{lj}=1} v_j\Big) \times E\Big(\sum_{j, b_{lj}=1} w_j\Big)\Big) \\
&\stackrel{\star}{=} \Big(\Big(\prod_{j, a_{1j}=1} E(v_j)\Big) \times \Big(\prod_{j, b_{1j}=1} E(w_j)\Big), \ldots, \Big(\prod_{j, a_{lj}=1} E(v_j)\Big) \times \Big(\prod_{j, b_{lj}=1} E(w_j)\Big)\Big) \\
&= \Big(\prod_{j, a_{1j}=1} E(v_j), \ldots, \prod_{j, a_{lj}=1} E(v_j)\Big) \odot \Big(\prod_{j, b_{1j}=1} E(w_j), \ldots, \prod_{j, b_{lj}=1} E(w_j)\Big) \\
&\stackrel{\star}{=} \Big(E\Big(\sum_{j, a_{1j}=1} v_j\Big), \ldots, E\Big(\sum_{j, a_{lj}=1} v_j\Big)\Big) \odot \Big(E\Big(\sum_{j, b_{1j}=1} w_j\Big), \ldots, E\Big(\sum_{j, b_{lj}=1} w_j\Big)\Big) \\
&= E\Big(\sum_{j, a_{1j}=1} v_j, \ldots, \sum_{j, a_{lj}=1} v_j\Big) \odot E\Big(\sum_{j, b_{1j}=1} w_j, \ldots, \sum_{j, b_{lj}=1} w_j\Big) \\
&= E(Av) \odot E(Bw) \\
&= E\big(\mathscr{E}(m_0)\big) \odot E\big(\mathscr{E}(m_1)\big) \\
&= EV(m_0) \odot EV(m_1),
\end{aligned}
$$

where the equalities $\star$ follow from the additive homomorphic property of the GM encryption function E, and $\odot$ denotes the componentwise multiplication of two vectors in $\mathbb{Z}_2^l$ (Hadamard product). Moreover, the equality $E(v + w) = E(Av + Bw)$ is nothing but the randomization of v and w by Lemma 20. Hence

$$
\mathscr{D}(D(EV(m_0 \times m_1))) = \mathscr{D}(D(EV(m_0) \odot EV(m_1))).
$$

Re-randomization is possible and required for the SYY algorithm to prevent leakage of any substantial information about the encryptions. Let $r_1,\ldots,r_l \in \mathbb{Z}_n^*$ be chosen uniformly at random and $c = EV(m)$. Then

$$
\begin{aligned}
&({r_1}^2,\ldots,{r_l}^2) \odot EV(m) = \\
&\quad = ({r_1}^2,\ldots,{r_l}^2) \odot (E(v_1),\ldots,E(v_l)) \\
&\quad = ({r_1}^2,\ldots,{r_l}^2) \odot ({y_1}^2 \times x^{v_1} \pmod{n},\ldots,{y_l}^2 \times x^{v_l} \pmod{n}) \\
&\quad = ({r_1}^2 \times {y_1}^2 \times x^{v_1} \pmod{n},\ldots,{r_l}^2 \times {y_l}^2 \times x^{v_l} \pmod{n}) \\
&\quad = ((r_1 \times y_1)^2 \times x^{v_1} \pmod{n},\ldots,(r_l \times y_l)^2 \times x^{v_l} \pmod{n})
\end{aligned}
$$

which is also a valid encryption of m. Hence

$$\mathscr{D}(D(({r_1}^2,\ldots,{r_l}^2) \odot c)) = \mathscr{D}(D(c)).$$

12.5 Security

The semantic security of the Sander-Young-Yung cryptosystem follows directly from the semantic security of the Goldwasser-Micali cryptosystem, which is based on the intractability of the QR[n] problem.

12.6 Example

Let $m = 0$ and $l = 4$. Then,

$$EV(0) = E(\mathscr{E}(0)) = E(1,0,1,0) = (c_1,c_2,c_3,c_4) = c,$$

where $m = 0$ is encoded as the nonzero vector $v = (1,0,1,0)$. Here we apply bit-by-bit GM encryption.

Likewise c is decrypted by using GM decryption and then decoded by $\mathscr{D}$.

$$\mathscr{D}(D(c)) = \mathscr{D}(1,0,1,0) = 0 = m.$$

Re-randomization of messages works bit by bit like GM re-randomization. For further information on GM encryption, decryption and re-randomization, we refer the reader to Chapter 4.

To exemplify the homomorphic property, let $m' = 0$ and let m' be also encoded as $v = (1,0,1,0)$. Consider the following 4×4 invertible matrices over $\mathbb{Z}_2$ for randomization process.

$$A = \begin{bmatrix} 1\,0\,0\,0 \\ 1\,1\,0\,0 \\ 1\,1\,1\,0 \\ 1\,1\,1\,1 \end{bmatrix}, \quad B = \begin{bmatrix} 1\,0\,0\,0 \\ 0\,1\,0\,0 \\ 1\,0\,1\,0 \\ 0\,0\,0\,1 \end{bmatrix}$$

Then we have

$$\begin{aligned} EV(m) \odot EV(m') &= E(\mathscr{E}(m)) \odot E(\mathscr{E}(m')) \\ &= E(Av) \odot E(Bv) \\ &= E(1,1,0,0) \odot E(1,0,0,0) \\ &\stackrel{\star}{=} E((1,1,0,0) \oplus (1,0,0,0)) \\ &= E(0,1,0,0), \end{aligned}$$

where the equality $\star$ follows from the homomorphic property of GM scheme.

Clearly, $\mathscr{D}(D(EV(m) \odot EV(m'))) = \mathscr{D}(0,1,0,0) = 0 = m \times m' = \mathscr{D}(D(EV(m \times m')))$, which means $EV(m) \odot EV(m') = EV(m \times m')$.

References

1. Benaloh J.C.: Verifiable Secret-Ballot Elections. PhD Thesis, Yale University (1988)
2. Benaloh J.C.: Dense probabilistic encryption. In: Selected Areas in Cryptography (SAC), pp. 120–128. (1994)
3. Boneh D., Venkatesan R.: Breaking RSA may not be equivalent to factoring. In: Nyberg K. (ed) Advances in Cryptology — EUROCRYPT'98, LNCS 1403, pp. 59–71. Springer, Berlin, Heidelberg (1998)
4. Boneh D., Goh E., Nissim K.: Evaluating 2-DNF formulas on ciphertexts. In: Kilian J. (ed) Theory of Cryptography, LNCS 3378, pp. 325–341. Springer, Berlin, Heidelberg (2005)
5. Damgård I., Jurik M.: A generalisation, a simplification and some applications of Paillier's probabilistic public-key system. In: Kim K. (ed) Public Key Cryptography. PKC 2001. LNCS 1992, pp.119–136. Springer, Berlin, Heidelberg (2001)
6. ElGamal T.: A public key cryptosystem and a signature scheme based on discrete logarithms. In: Blakley G.R., Chaum D. (eds) Advances in Cryptology-CRYPTO '84, LNCS 196, pp. 10–18. Springer, New York, USA (1985)
7. Fousse L., Lafourcade P., Alnuaimi M.: Benaloh's dense probabilistic encryption revisited. In: Nitaj A., Pointcheval D. (eds) Progress in Cryptology — AFRICACRYPT 2011. LNCS 6737, pp. 348–362. Springer, Berlin, Heidelberg (2011)
8. Gentry C.: A Fully Homomorphic Encryption Function. PhD Thesis, Stanford University (2009)
9. Goldwasser S., Micali S.: Probabilistic encryption. J. Comput. Syst. Sci., 28(2), pp. 270-299 (1984)
10. Koç Ç.K.: editor. Cryptographic Engineering. Springer (2007)
11. Lynn B.: On the Implementation of Pairing-Based Cryptosystems. PhD thesis, Stanford University (2007)
12. Maas M.: Pairing-Based Cryptography. MSc thesis, Technische Universiteit Eindhoven (2004)
13. Meffert D.: Bilinear Pairings in Cryptography. MSc thesis, Radboud Universiteit (2009)
14. Menezes A.J., van Oorschot P.C., Vanstone S.A.: Handbook of Applied Cryptography. CRC Press (1997)
15. Micali S., Rackoff C., Sloan B.: The notion of security for probabilistic cryptosystems. SIAM J. Comput., 17(2), pp.412–426 (1988)
16. Miller V.: The Weil pairing, and its efficient calculation. J. Cryptol., 17(4), pp. 235–261 (2004)
17. Naccache D., Stern J.: A new public key cryptosystem based on higher residues. In: Proceedings of the 5th ACM Conference on Computer and Communications Security, pp.59–66 (1998)
18. Okamoto T., Uchiyama S.: A new public-key cryptosystem as secure as factoring. In: Nyberg K. (ed) Advances in Cryptology — EUROCRYPT'98. LNCS 1403, pp. 308–318. Springer, Berlin, Heidelberg (1998)
19. Paillier P.: Public-key cryptosystems based on composite degree residuosity classes. In: Stern J. (ed) Advances in Cryptology. EUROCRYPT 1999. LNCS 1592, pp. 223–238. Springer, Berlin, Heidelberg (1999)
20. Pollard J.M.: Monte Carlo methods for index computation $\pmod{p}$. Math. Comput., 32(143), pp. 918–924 (1978)
21. Rivest R., Shamir A., Adleman L.: A method for obtaining digital signatures and public-key cryptosystems. Commun. ACM, 21(2), pp. 120–126 (1978)
22. Rivest R., Adleman L., Dertouzos M.: On data banks and privacy homomorphisms. In: DeMillo R.A., Dopkin D.P., Jones A.K., Lipton R.J. (eds) Foundations of Secure Computation, pp 169–180. Academic Press (1978)
23. Rosen K.H.: Elementary Number Theory and Its Applications. 6th Edition, Addison-Wesley (2011)
24. Sander T., Young A., Yung M.: Non-interactive cryptocomputing for NC. In: FOCS 1999, IEEE, pp. 554–567 (1999)

Ç. K. Koç et al., *Partially Homomorphic Encryption*,
https://doi.org/10.1007/978-3-030-87629-6

25. Stein J.: Computational problems associated with Racah algebra. J. Comput. Phys., 1(3), pp. 397–405 (1967)
26. Stipčevič M., Koç Ç.K.: True random number generators. In: Koç Ç.K. (ed) Open Problems in Mathematics and Computational Science, pp. 275–315, Springer, December 2014 (2014)
27. Washington L.C.: Elliptic Curves. Chapman & Hall/CRC (2003)
28. Weil A.: Sur les fonctions algébraiques à corps de constantes finis. C. R. Acad. Sci., 210(17), pp. 592–594 (1946)

Further Reading

29. Chen Y.Y.: The BGN Public-Key Cryptosystem and Its Application to Authentication, Oblivious Transfers, and Proof-of-Visit. PhD Thesis, Chinese University of Hong Kong (2006)
30. Damgård I., Jurik M., Nielsen J.B.: A generalization of Paillier's public-key system with applications to electronic voting. Int. J. Inf. Secur., 9(6), pp. 371–385 (2010)
31. El Mrabet N., Joye M.: Guide to Pairing-Based Cryptography. Chapman & Hall/CRC (2017)
32. Fontaine C., Galand F.: A survey on homomorphic encryption for nonspecialists. EURASIP J. Inf. Secur., Article number: 013801 (2007)
33. Henry K.: The Theory and Applications of Homomorphic Cryptography. MSc Thesis, Computer Science, University of Waterloo, Canada (2008)
34. Jurik M.: Extensions to the Paillier Cryptosystem with Applications to Cryptological Protocols. PhD Thesis, University of Aarhus (2004)
35. Koblitz N.: A Course in Number Theory and Cryptography. Springer-Verlag (1994)
36. Koç Ç.K.: High-speed RSA implementation. Technical Report TR 201, RSA Laboratories, 73 pages, November 1994 (1994)
37. O'Keeffe M.: The Paillier cryptosystem, A look into the cryptosystem and its potential application. College of New Jersey (2008)
38. Paar C., Pelzl J.: Understanding Cryptography. Springer-Verlag (2010)
39. Pettersen N.: Applications of Paillier's Cryptosystem. MSc Thesis, Norwegian University of Science and Technology (2016)
40. Rodríguez-Henríquez F., Saqib N.A., Pérez A.D., Koç Ç.K.: Cryptographic Algorithms on Reconfigurable Hardware. Springer (2007)
41. Yamamura A., Jaycayova T., Kurokawa T.: Homomorphic encryption functions and cryptographic protocols. Kyoto Univ. Dep. Bull. Pap., 1437, pp. 97–106 (2005)
42. Yi X., Paulet R., Bertino E.: Homomorphic encryption. In: Homomorphic Encryption and Applications. Springer Briefs Comput. Sci., pp. 27–46. Springer (2014)

Index

Ç. K. Koç et al., *Partially Homomorphic Encryption*,
https://doi.org/10.1007/978-3-030-87629-6

Printed in the United States
by Baker & Taylor Publisher Services